ESSENTIAL EARTH IMAGING

FOR GIS

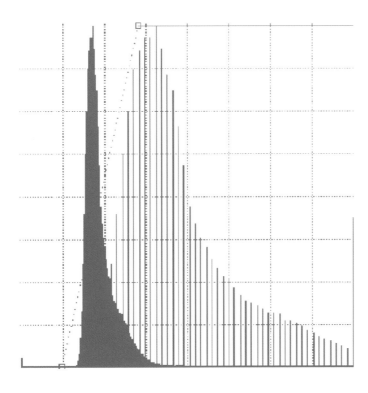

Lawrence Fox III

Esri Press
REDLANDS | CALIFORNIA

Cover image from Aleksander Todorovic/Shutterstock.com.

Esri Press, 380 New York Street, Redlands, California 92373–8100
Copyright © 2015 Esri
All rights reserved. First edition 2015

Printed in the United States of America
19 18 17 16 15   1 2 3 4 5 6 7 8 9 10

Library of Congress Cataloging-in-Publication Data

Fox, Lawrence, 1947-
  Essential Earth imaging for GIS / Lawrence Fox, III. — First edition.
    pages cm
  Includes bibliographical references and index.
  ISBN 978-1-58948-345-3 (pbk. : alk. paper) — ISBN 978-1-58948-431-3 (electronic)
  1. Earth sciences—Remote sensing.   2. Geographic information systems.   3. Image analysis.   I. Title.
  G70.4.F65 2015
  910.285—dc23   2015010934

Ask for Esri Press titles at your local bookstore or order by calling 800–447–9778, or shop online at esri.com/esripress. Outside the United States, contact your local Esri distributor or shop online at eurospanbookstore.com/esri.

Esri Press titles are distributed to the trade by the following:

In North America:
Ingram Publisher Services
Toll-free telephone: 800–648–3104
Toll-free fax: 800–838–1149
E-mail: customerservice@ingrampublisherservices.com

In the United Kingdom, Europe, Middle East and Africa, Asia, and Australia:
Eurospan Group
3 Henrietta Street
London WC2E 8LU
United Kingdom
Telephone: 44(0) 1767 604972
Fax: 44(0) 1767 601640
E-mail: eurospan@turpin-distribution.com

Contents

Introduction

As a geographic information system (GIS) professional or student, you are probably familiar with the simple display of aerial images. Aerial images have been used as backdrops in GIS for many years. However, digital imaging technology is advancing. You now have access to georeferenced Earth imagery from multiple sources and with multiple wavelength bands. The image processing functionality of your GIS software is keeping pace with these advancements. Because of these advancements, you now need a greater knowledge of sophisticated Earth imaging (Esri 2014a; Intergraph 2014; GRASS GIS 2014). This knowledge should include how to use software tools to display, coregister, and enhance Earth imagery within GIS. *Essential Earth Imaging for GIS* can help you attain that knowledge.

Essential Earth Imaging for GIS provides a basic education in imaging technology and management, promoting the effective use of imaging tools. Undergraduate students, particularly those with little or no experience using Earth imagery, will benefit from using this book as a reference for introductory GIS courses that include multispectral image display and analysis. Students and professionals can use this book as a general Earth imagery reference and also as a reference for the specific Earth imagery topics listed in the table of contents. This book includes concepts and methods of image formation and manipulation that enable you to efficiently and effectively display, coregister, enhance, interpret, and delimit features from an image.

Except for the brief history of Earth imaging provided in chapter 1, the book contains discussions organized by the image data type and the method of image formation rather than a chronological description of imaging technologies. This was done in the belief that GIS professionals and students are interested in data produced by all of the sensors producing that data type, regardless of when they were developed. For example, a digital elevation model produced from a laser sensor (perfected in the last 10 years) might resemble an elevation model produced from a stereo pair of aerial photographs (developed in the 1940s). Or, the two elevation models might look very different because of surface conditions. Therefore, imaging technologies are grouped by the kind of image produced and by the technology's attributes: two-dimensional images or three-dimensional datasets. The presentation of imaging technologies is also organized by how the technologies work—using active or passive sensors—and by what type of electromagnetic radiation the technologies use.

Chapter 1 provides a rationale for incorporating Earth imagery into GIS, some historical background on the development of Earth imaging, a description of the unique raster structure of a digital image, and a summary of the different types of imagery used in GIS. The physical basis and methods of image formation are described in chapter 2, and the effects of the atmosphere on image quality are explained in chapter 3. A user's perspective on the workings of two-dimensional imaging sensors and the attributes of the images they produce is provided in chapter 4. The display and enhancement of multispectral imagery in gray scale, natural color, and false color is explained in chapter 5.

Chapters 6 and 7 provide an introduction to aspects of Earth imaging that have traditionally been the purview of imaging specialists. Chapter 6 includes an introduction to photogrammetry that will enhance the effective use of imaging GIS software for compiling image mosaics and measuring image features. Chapter 6 also includes information and references about generating elevation models using photogrammetry, which requires specialized software and further study. The creation of point clouds with lidar sensors is also explained conceptually in chapter 6. Automated processing of point cloud data to create digital surface models is introduced, salient points explained, and further reading suggested. Digital elevation models generated by interferometric radar sensors are described only briefly in chapter 6 because GIS professionals normally work with the digital models created from such advanced technology rather than processing the interferometric images.

Image processing is described in chapter 7. Image providers increasingly process images to improve image quality, register images to map projections, and reduce atmospheric effects. While traditionally beyond the scope of normal practice for GIS professionals, some of these processing functions are being incorporated into GIS software. Chapter 7 will familiarize you with processing steps and provide guidance as to how much, and in what order, to process an image before incorporating it into GIS. This information will also help you understand how images are processed by image providers before they are delivered.

Earth images are often used within a GIS to create new geographic information or to update existing information, such as adding new roads or land-use classes to existing vector map layers. Maps, engineering drawings, and three-dimensional vector models of existing structures and landscapes can be extracted from imagery or from point clouds. Thus the GIS itself becomes an effective tool for creating new information that, in turn, adds to the usefulness of the GIS for further geographic analysis. Chapter 8 provides a detailed description of visual image interpretation techniques and methods for manually delineating cartographic features on an image using GIS software. It also provides an introduction to automated methods of image classification and feature extraction. Although some automated image classification tools are increasingly available in GIS software, the most sophisticated algorithms require specialized image processing software and further study of references included in this chapter.

Companion exercises that use Esri's ArcGIS for Desktop software to examine Earth imagery are available on the Esri Press "Book Resources" web page, http://esripress.esri.com/bookresources. Exercise data and access to a 180-day trial of ArcGIS software are also available on the "Book Resources" web page. Click the appropriate book title, and then click the links under "Resources" to download the exercise PDF and the data, and access the trial software. You will need the authorization code printed on the inside back cover of this book to access the trial software. The exercises are designed to enhance learning after reading specific chapters in this book. Instructions are included with the exercises.

Chapter 1
Overview of imaging GIS

Imaging GIS, a term used in the medical imaging community (Wang 2012), is adopted here to describe a geographic information system (GIS) that displays, enhances, and facilitates the analysis of Earth images from satellites, aircraft, and, most recently, unmanned aerial systems (UAS). Earth imaging has roots in aerial photogrammetry, which is the "science or art of obtaining reliable measurements by means of photography" (American Society of Photogrammetry 1952). Photogrammetry expanded in the 1960s to include images from the thermal infrared and microwave regions of the electromagnetic spectrum. Orbiting satellites were also being considered as platforms for imaging sensors during these years. As a result, the term *remote sensing* was coined to replace the more limiting terms *aerial* and *photograph* and embrace these developments (Colwell 1960). Remote sensing continued to develop in the 1970s and 1980s with active, microwave radar, which greatly expanded the range of wavelengths used to create images (Slama, Theurer, and Henriksen 1980). More recently, laser scanners (lidar systems) have been employed to create three-dimensional point clouds, a very different kind of "image" that facilitates three-dimensional modeling of the Earth's surface (Renslow 2012).

Earth images provide photo-realistic representations of the Earth's surface that increase the interpretability and usefulness of GIS in unique ways. The human eye recognizes many surface features from the wealth of detail present in Earth images, often more easily than from cartographic representations of those features. For example, maps often contain only building footprint symbols for a single-family residential house. An image often contains sufficient detail to not only recognize the building as a single-family residence, but enough to assess the approximate value of the property by inference from landscaping, driveway type, and other clues. Figure 1-1 shows the difference in the level of detail that can exist between an aerial photo and a computer-generated map.

Figure 1-1. Two views of Fashion Island shopping center in Newport Beach, California, in 2009. The image on the left is an aerial photograph, and the image on the right is a typical computer-generated map.

Map courtesy of US Geological Survey. Topographic Mapping Program provided by Esri ArcGIS Online; Esri, DeLorme, NAVTEQ, USGS, USDA, EPA. Image courtesy of National Agriculture Imagery Program, US Department of Agriculture, Farm Service Agency, California Department of Fish and Wildlife, Biogeographic Data Branch.

The myriad recognizable features in the photograph in figure 1-1 would add to the visual information of any map layer in which the photograph was used. In addition, features not shown on the map in figure 1-1 could be identified and delineated from the photograph using a GIS, which creates even more information.

Structure of a two-dimensional digital image

Images are traditionally displayed as raster layers in a GIS, a data structure that most GIS professionals understand. However, images can be more complex than other types of raster data, which can make it challenging for those without remote sensing training to effectively display and enhance them. To review, the raster data structure consists of grid cells arranged in rows and columns. Each cell has associated attributes. The rows and columns provide a coordinate system to locate any grid cell. These image coordinates may be coregistered with a number of Earth coordinate systems such as latitude/longitude, universal transverse Mercator (UTM), or state plane coordinates, as appropriate for specific applications.

The grid cells in an image raster represent picture elements (pixels) of a specific size and shape on the ground (normally square). The key difference between an image raster and other raster grids (such as a digital elevation model) is that the cell attributes of an image are not associated with any specific descriptive variable such as elevation or land cover. The cell values in an image represent brightness. These cell values have a large range of levels (traditionally 256 levels, but recently ranges of up to 2,048 levels have been introduced). Brightness is more analogous to a continuous variable than a categorical variable because it can contain so many levels.

The brightness values contained within the grid cells of an image are seldom displayed as numbers. Rather, shades of gray are sometimes applied by the GIS software to form the classic black-and-white (grayscale) image display (figure 1-2).

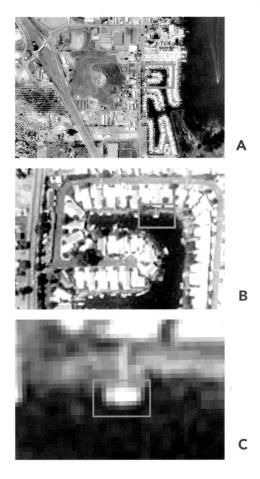

A

B

C

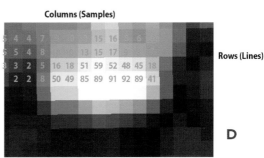

D

Figure 1-2. The raster data structure of an image showing the relationship between numerical brightness values and shades of gray in a simple grayscale rendition. *A*, A digital image of a small boat harbor is shown. *B*, An enlargement of the area in the rectangle is shown. *C*, An enlargement of one pier and floating dock shows individual pixels of 1 × 1 meter on the ground. *D*, An enlargement of the end of the dock displays the digital brightness values of the pixels as numbers.

Photograph courtesy of US Geological Survey, DOQQ, Clear Lake, California, 1993.

The numerical brightness value of any single pixel is usually not informative to the image interpreter. However, the interpreter can use the geographic pattern of relative brightness values to help correctly interpret Earth features that are much larger than a single pixel. The exception is *quantitative remote sensing*, in which environmental variables (such as surface temperature) are inferred from upwelling radiation (further explained in chapter 2).

Another factor that makes image data more complicated than other raster layers is color. Digital color photographs are managed in most general-purpose computer systems as one file in familiar formats (such as picture.tif or picture.jpg). As a result, the casual user may be unaware of the tristimulus theory of color vision, which states that all colors can be formed from various shades of three additive primary colors: red, green, and blue. Every color image is actually three images superimposed in various shades of those three colors (explained in detail in chapter 5). The information contained within color images is best exploited when the three primary additive colors are separated into three raster layers. Images are displayed as three separate layers in most GIS software packages rather than as one layer in common image formats. This three-layer display allows the analyst to manipulate each additive color separately, increasing interpretability of the image. Advanced sensors provide wavelength bands in addition to the visual range, thus providing more information about the Earth's surface than the spectral sensitivity of consumer-grade color photography. The most information is extracted from imagery when false-color combinations are used in which invisible radiation is made visible on screen (Campbell and Wynne 2011). This process is further explained in chapter 5.

Three-dimensional data

GIS software is increasingly able to produce realistic three-dimensional visualizations of terrain from digital elevation models. These perspective views are enhanced when imagery is displayed as an overlay on an elevation model (figure 1-3).

Pairs of aerial images acquired in an overlapping flight pattern can be used to create elevation models through the measurement processes of photogrammetry. Although traditionally the domain of specialized image processing software, photogrammetric functionality is increasingly being incorporated into GIS (Esri 2014a; further explained in chapter 6).

Figure 1-3. A grayscale image draped over a digital elevation model of Sausalito, California, in 1991.

Image and elevation data courtesy of US Geological Survey. Visualization created by the author.

Three-dimensional data may also be represented by *point clouds* in some GIS software. Each data point has three attributes in this data model: an x-coordinate and a y-coordinate for the horizontal position and a z-coordinate for the vertical position. Point cloud data may be displayed in perspective view with visualization software embedded in GIS (figure 1-4). The visualizations provide a reasonably realistic representation of the scene, depending on the number of points in the cloud.

Figure 1-4. An example of point cloud data for New York City, color-coded by elevation.

New York City lidar image courtesy of The Sanborn Mapping Company, Inc. Data acquired and created through a sustainable City University of New York (CUNY) project on behalf of the NYC Solar America City Partnership. Used by permission.

The generation of point clouds has been facilitated by the recent development of laser scanning devices (lidar) that provide millions of geographically registered points very rapidly (Vosselman and Maas 2010). The point cloud structure provides detailed information using very large amounts of data, compared with traditional two-dimensional raster grids. GIS software is rapidly being developed to incorporate point cloud data for the visualization and detailed measurement of features that range from individual structures to entire landscapes (further explained in chapter 6).

Essential Earth Imaging for GIS

Chapter 2
The physical basis and general methods of remote sensing

Knowing how images are formed by sensors will help you effectively use the images created by them. In this chapter, you will learn the principles of electromagnetic (EM) radiation, the engine of image formation in virtually all Earth imaging systems. Underwater acoustic sonar is a well-known example of using sound to generate images, but this form of remote sensing requires specialized software for image display, beyond GIS. Knowing what EM radiation is and how it interacts with the environment will help you interpret the variety of digital image and elevation data available today and planned for tomorrow. In this chapter, you will also learn about current imaging systems and the capabilities and limitations of various remote sensing methods. You will learn about aerial and spaceborne platforms and how their characteristics influence the attributes of the imagery collected from them. Further detail on how sensors generate images is provided in chapter 4 for two-dimensional images and in chapter 6 for three-dimensional point clouds.

How remote sensing works

In order to remotely sense features on the Earth's surface, information must be conveyed from the feature to the sensor. Information about the size, shape, texture, color (or tone), and sometimes the temperature of a feature is conveyed to the sensor by EM radiation. The EM radiation can come from the sun or the Earth and be detected passively by the sensor. The EM radiation can also be generated by the sensor itself to accomplish active remote sensing.

The diagram in figure 2-1, the energy flow profile, illustrates the process of image formation and how remote sensing works. The profile has three components that indicate the flow of EM radiation: the source, the scene, and the sensor. Figure 2-1 illustrates the profile of energy flow when the source of the EM radiation is the sun and the radiation reflects from an object in the scene to be sensed passively by the sensor. Aerial photography and most satellite sensing are done in this way with optical instruments. The diagram is modified to show passive thermal remote sensing and active microwave remote sensing later is this chapter.

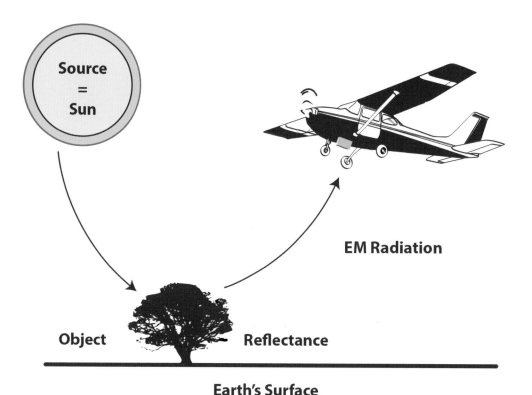

Figure 2-1. The energy flow profile shows the sensor at some distance from the objects of interest. The sun is the source of electromagnetic radiation that creates the image.

Remote sensing precludes direct (in situ) measurement by field observation. However, a percentage of remotely sensed information is commonly evaluated on the ground using a scientific sampling plan, a procedure called *ground truthing*.

Electromagnetic radiation

Remote sensing systems use EM radiation to form images of Earth features. To understand the remote sensing process, you must know a little about EM radiation.

All matter with a temperature above absolute zero produces EM radiation (Waldman 1983). EM radiation travels at a rapid and constant velocity: the speed of light, or 186,000 miles per second. Some experiments show that EM radiation has wave properties, and others show that it behaves as if it were composed of particles (photons). Both the wave and particle nature of EM

Essential Earth Imaging for GIS

radiation are helpful to understanding image formation. Waves allow the separation of radiation into separate wavelength intervals (wavelength bands), which adds greatly to the information content of images. The particle nature of EM radiation provides the energy that forms the image.

Remote sensing is accomplished with EM radiation emitted by the sun (light) and by the Earth (heat). The sun is extremely hot (6,000°C at the surface) and emits tremendous amounts of light (400 trillion terawatts at the sun's surface, about 1,360 watts per square meter arriving at the Earth). The radiation is produced over a range of wavelengths between about 0.1 and 3.0 micrometers, with a peak wavelength of about 0.5 micrometers (figure 2-2).

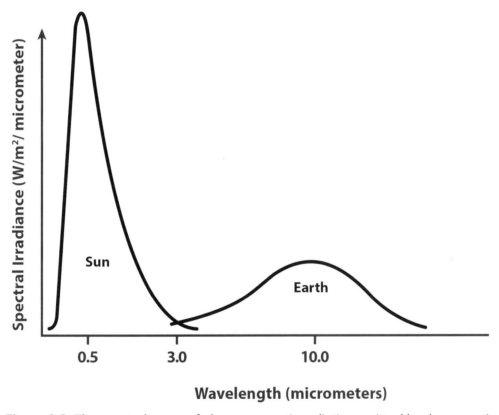

Figure 2-2. The spectral range of electromagnetic radiation emitted by the sun as it arrives at the top of Earth's atmosphere and the spectral range of radiation emitted by the Earth itself.

The Earth, by contrast, is cool (average surface temperature of about 15°C) and emits smaller amounts (an average of about 235 watts per square meter) of heat radiation over a broader range of longer wavelengths. The Earth radiates between about 3.0 and 50 micrometers, with a peak wavelength of emission of about 10 micrometers (figure 2-2).

Both short-wavelength solar radiation and long-wavelength Earth radiation are used to create images with very different information content. Images generated from reflected solar radiation provide information about the reflective properties of various Earth surfaces, allowing the discrimination of surface condition. Images generated from thermal radiation, emitted from the Earth, provide information about the relative temperatures of various surfaces. Most remote sensing is accomplished with EM radiation produced by the sun, including the most common form, aerial photography.

The term *light* usually describes only a narrow band (0.4–0.7 micrometers) of the entire EM wavelength spectrum, the band to which our eyes are sensitive and the band in which most of the sun's radiation is emitted. Visible light has been used throughout the history of remote sensing to capture images, first in shades of gray and then in color. However, the most information-laden images are captured with a combination of visible and nonvisible wavelengths.

The particles (photons) of EM radiation provide the energy that creates the image. The wave and particle nature of EM radiation are inversely related. For example, short wavelengths carry high energy. High-energy radiation facilitates image formation because these strong signals easily overcome system noise and atmospheric interference (chapter 3) to allow finely detailed images to be collected. Short-wavelength radiation can be captured with simpler imaging systems (the common camera, for example) than long-wavelength radiation, which requires more sophisticated sensors. And yet, even though they are captured by more sophisticated sensors, images formed by long wavelengths (thermal images, for example) are often of poorer quality because of the low energy content.

Passive remote sensing

Remote sensing in which the sensor reacts to radiation produced elsewhere in the environment is called *passive remote sensing* because the sensor does not generate its own radiation. Passive remote sensing systems can detect different wavelengths of the EM radiation spectrum. Earth features reflect different parts of the spectrum at different rates. This reflectance percentage is the key to differentiating features on the Earth when using solar radiation (light) for passive remote sensing.

Reflectance patterns and the spectral signature

At the instant an image is formed, the downwelling radiation from the sun is virtually constant across the image area. The relative amount of radiation reflected from the Earth's various surfaces produces different brightness values in the resulting image, thus differentiating features on the ground. In addition to various Earth surfaces being generally brighter or darker across the entire spectrum, specific surfaces reflect different amounts of solar radiation in different wavelength intervals (figure 2-3).

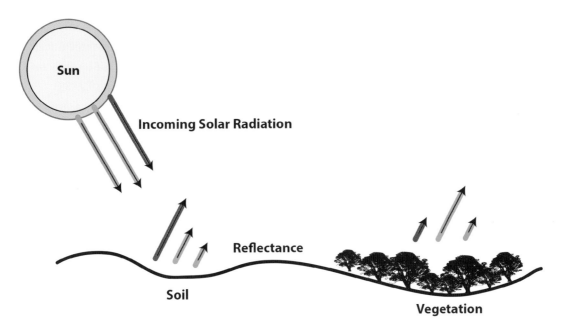

Figure 2-3. Spectral reflectance of soil and vegetation in the visible wavelengths that human eyes perceive as blue, green, and red light.

Figure 2-3 provides a technical answer to the question of why living plants appear green to human eyes and soil often appears brown. The lengths of the arrows in the diagram indicate the relative amount of sunlight incoming and reflected from the two surfaces at different wavelengths. The green plants reflect more light in the middle portion of the visible spectrum, those wavelengths that humans perceive as green. Because of chlorophyll absorption, the plants reflect less light in the short and long end of the visible spectrum, those wavelengths that humans perceive as blue and red. The soil reflects more light in the long wavelengths of the visible spectrum than in the short wavelengths, causing human eyes to perceive a reddish-brown color. Further details on color theory and perception are presented in chapter 5.

This pattern of reflectance values across multiple wavelength bands provides practitioners with the diagnostic data they need to identify features on the ground. Remote sensing practitioners call this critical concept the *spectral signature*. It is important to consider the spectral signature when working with invisible wavelengths because we do not know how surfaces would appear if human vision was possible in those wavelengths. The advantage of including invisible wavelengths in remote sensing is that some surfaces have different reflectance in invisible wavelengths than in visible ones. Including both visible and invisible wavelengths increases the diagnostic power of remote sensing.

EM radiation with wavelengths between 0.7 and about 50 micrometers is commonly called *infrared radiation* by physicists. The term *infrared* is commonly used in casual, nontechnical conversation by most people to designate heat radiation produced by the Earth or by objects on the Earth. However, the infrared region includes radiation emitted by the sun (0.7–3.0 micrometers) and radiation emitted by the Earth (3.0–50 micrometers), as shown in figure 2-2. As a result, confusion exists as to the ability of infrared-sensitive films, or of infrared-sensitive digital cameras, to sense heat. Photographic films and digital cameras with sensitivities out to 1.3 micrometers cannot sense heat. This spectral range is clearly measuring reflected EM radiation (below 3.0 micrometers) that is produced by the sun and reflected from the Earth's surface. Those working in remote sensing commonly define a reflected infrared portion of the EM spectrum of 0.7–3.0 micrometers to clear up this confusion. This reflected infrared radiation has nothing to do with the temperature of the Earth's surface, rather the sensor measures the amount of solar radiation reflected from the surface when operating in the spectral region between 0.7 and 3.0 micrometers.

EM radiation wavelengths detected by passive sensors

Further divisions of the reflected EM radiation spectrum (visible plus reflected infrared) are used in remote sensing because many sensors are filtered to receive multiple narrow spectral bands in the visible (0.4–0.7 micrometers) and in the reflected infrared (0.7–3.0 micrometers) ranges. The visible range is traditionally divided into three bands: blue (0.4–0.5 micrometers), green (0.5–0.6 micrometers), and red (0.6–0.7 micrometers). Many sensors also include a near-infrared (NIR; 0.7–1.3 micrometers) band. Advanced sensors may even include one or two short-wave infrared (SWIR; 1.3–3.0 micrometers) bands.

Figure 2-4 illustrates the relationship between wavelength regions and reflectance percentages of three common Earth surfaces: vegetation, soil, and water.

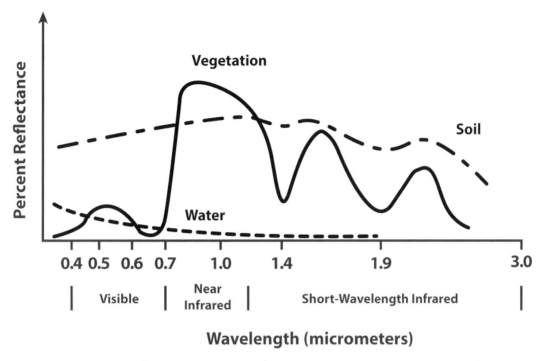

Figure 2-4. Spectral reflectance signatures for vegetation (solid line), soil (dash-dot line), and water (dash line) in the visible and reflected infrared spectrum.

The graph in figure 2-4 is an abstraction of figure 2-3. Instead of variable-length arrows showing relative amounts of upwelling radiation, this figure indicates percentage reflectance from three surfaces. A higher percentage reflectance produces more upwelling radiation from the surface, which creates brighter pixels at the sensor. Therefore, the height of the lines in figure 2-4 determines the relative brightness of the feature, not considering atmospheric effects (chapter 3).

Soil is fairly bright throughout the graphed wavelengths. It is slightly higher in the visible red band (0.6–0.7 micrometers) than in the shorter portions of the visible spectrum (0.4–0.6 micrometers). Vegetation is brighter in the visible green band (0.5–0.6 micrometers) and especially bright in the NIR band (0.7–1.3 micrometers) than at other wavelengths. Water is very dark at wavelengths longer than 0.7 micrometers.

These relationships show how you can use information from multiple spectral bands to identify and differentiate features of interest in the resulting images. For example, water can be easily differentiated from soil and vegetation by sensing in the NIR band, where water is very dark and vegetation and soil are both bright in the resulting image. Vegetation and soil are hard to distinguish from each other in the NIR band, but vegetation can be easily separated from soil in the visible red band (0.6–0.7 micrometers), where vegetation and water are both much darker than soil. So, the NIR band can be used to separate the water from the other two surfaces, and the visible red band can be used to separate the soil from the other two surfaces. What is left must be vegetation. All three features can be separated from each other using both bands in combination. This is the central thesis of multispectral remote sensing. More information on the exploitation of spectral reflectance differences using multispectral sensors is provided in chapter 4.

Spectral differentiation of features on the ground depends on consistent reflectance patterns regardless of viewing angle and sun angle. Fortunately for users of remote sensing, viewing angles are normally near vertical with images acquired at near midday. This minimizes differences in both viewing angle and sun angle. In addition, most natural features reflect radiation in a diffuse manner. That is, as long as the sun is high in the sky, a view angle slightly departing from vertical will not produce significantly lighter or darker features relative to the sun angle. Objects having some height above terrain do cast shadows, however, and when viewing into the sun or away from the sun brightness variations occur in Earth images containing vertical objects such as shrub/tree vegetation and walls/buildings. Interpreters need to be aware of these differences and account for them. For example, a north-facing mountain slope may be imaged darker than south-facing slopes even with identical vegetation cover.

The one large exception to diffuse reflectance in the natural world is water, which has such a smooth surface that mirror-like reflectance commonly occurs. Mirror-like reflectance produces a strong glint when looking into the sun. Technically speaking, the glint occurs when the angle of incidence is equal to the angle of reflectance. Because the sun is high in the sky and the look angle is near vertical for most remote sensing operations, this water glint occurs quite often in near-vertically oriented images (figure 2-5).

Figure 2-5. A satellite image of the Hawaiian Islands from the visible spectrum, showing sun glint on the right side of the image.

MODIS satellite image from May 27, 2003, courtesy of NASA.

Passive thermal remote sensing

The energy flow profile shown in figure 2-1 describes a type of passive remote sensing in which EM radiation is emitted by the sun, reflected from the Earth, and formed into an image in the sensor: the *reflectance model* of remote sensing. In addition to using the reflectance model, it is also possible to filter the sensor to receive only long wavelengths of EM radiation (3.0 to about 50 micrometers) and create an image passively from heat radiation emitted by the Earth. This process is called *heat-sensitive thermal infrared imaging*: the *emittance model* of remote sensing (Quattrochi and Luvall 2004). A slightly modified energy flow profile, with an example image, illustrates passive thermal remote sensing in which the source of the thermal EM radiation is the Earth's surface (figure 2-6).

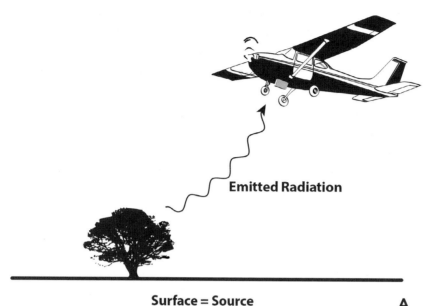

Emitted Radiation

Surface = Source

A

B

Figure 2-6. *A,* The energy flow profile for thermal infrared remote sensing. *B,* An example of a thermal infrared image showing a leaky steam pipe warming the soil, which appears in a lighter tone in the lower left-center of the image.

Thermal infrared remote sensing can be done at night without sunlight because the Earth produces the radiation being sensed. In fact, nighttime is ideal for passive thermal infrared sensing because the differential warming of the day has equalized and various surfaces have stable surface temperatures. Thermal infrared remote sensing can also be accomplished during daylight hours because the wavelengths produced by the Earth are 20 times longer than solar wavelengths (figure 2-2), so properly filtered sensors will not confuse the two types of radiation.

The brightness values in a thermal infrared image are correlated with the temperature of the Earth's surface or with the surface of objects in the scene. The physics of heat radiation favor temperature discrimination and sensing of relative temperature differences because a slight difference in temperature between surfaces produces a fourth-power (temperature to the fourth power) change in the amount of radiation emitted (Quattrochi and Luvall 2004). As long as the temperature variation is in the general range of ambient Earth temperatures, little change in peak wavelength occurs for slightly warmer or colder surfaces. Thus, one wavelength band of operation (normally 8–12 micrometers) serves for most thermal sensing of ambient Earth temperatures. A slightly shorter-wavelength thermal infrared band of 3–5 micrometers is used for hot objects (approximately 800°C) such as flowing lava and wildfires. This band is used because the peak emission of hot objects is in this shorter-wavelength interval than the peak for ambient temperature surfaces.

Traditionally, images formed by these thermal sensors show warmer surfaces as brighter and cooler surfaces as darker in a simple grayscale representation. More recently, images have been color-coded to improve interpretability (chapter 5). Common applications require the images to show only the differences between these relative temperatures, allowing the interpreter to identify and map warm people, warm vehicles, and warm or cool water. Imaging the relative difference in temperature is valuable for detecting and mapping poorly insulated sections of building roofs, locating warm water effluents, or detecting circulation patterns in tidal estuaries. The estuary application works because incoming tidal waters are usually colder than water remaining in the estuary from the previous tide. However, surface temperature can be inferred from the brightness values of the thermal image (often called *brightness temperature* in quantitative remote sensing [Liang 2004]) when the sensor is calibrated using surfaces of known temperature.

One complication with thermal infrared sensing is that the amount of energy emitted is influenced by the *emissivity* of the surface (Campbell and Wynne 2011). Emissivity is an efficiency factor used to account for the lower emissions of real surfaces that are less than the theoretical emission from a perfect radiator (referred to by physicists as a *black body*). Thus, the temperature inferred by thermal infrared remote sensing may be less than the actual surface temperature. However, the emissivity factor is nearly constant at about 0.95 for most natural features, thus producing fairly accurate inferred surface temperature values for a variety of surface conditions. Anomalies occur when imaging some manufactured surfaces, such as steel building roofs, which are very poor emitters. These surfaces are sensed to be much cooler than they actually are, a condition that can aid in their identification if that is the goal rather than measuring the temperature of the roof.

Active remote sensing

It is also possible to build sensors that generate their own EM radiation, thus sensing actively, as shown by the energy flow profile in figure 2-7.

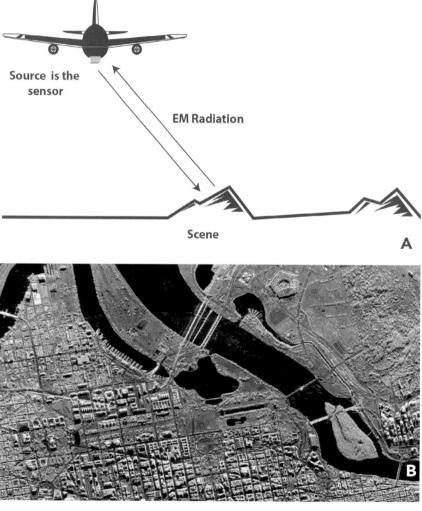

Figure 2-7. *A*, The energy flow profile for active sensing. *B*, An example radar image showing the Washington, DC, area.

Image courtesy of Sandia National Laboratories. Used by permission.

The source of the radiation is the sensor for active remote sensing. The wavelengths used and amounts of radiation can be tightly controlled when the sensor generates the radiation. This type of sensing is not dependent on available sunlight or thermal radiation emitted from the surface.

Active remote sensing can be accomplished with imaging *radar*, an acronym for radio detection and ranging. Radar sensors generate very long wavelength EM radiation (1–30 centimeters; that is, 10,000–300,000 micrometers). These very long waves penetrate clouds. In a strange twist of terminology, these longest wavelengths used in remote sensing are called *microwaves*. The term comes from the broadcasting industry in which these same wavelengths are the shortest waves used for television, relative to radio waves that are longer still (about 1–5 meters). Radar sensors generate a pulse of microwave radiation that travels from the sensor to the Earth's surface. Some of the pulse reflects back to the sensor, thus forming an image (explained in detail in chapter 4). Radar can acquire an image on the darkest rainy night, and is therefore preferred for military applications or any situation in which an image is needed now, without regard for the weather or time of day. Radar imaging has also been essential for civilian applications in very cloudy regions such as the tropical and polar regions of the Earth. Regardless of the need for all-weather sensing, radar provides a unique view of the Earth's surface because microwaves reflect very differently than short-wavelength solar radiation.

More recently, laser light has been used to actively create three-dimensional point clouds (figure 1-4). Lasers are active sensors that create their own short-wave radiation in the range of 0.5–2 micrometers. The three-dimensional point clouds produced by these sensors are developed using light detection and ranging (lidar) systems, explained further in chapter 6. The data are three dimensional and provide a very different view of the Earth than two-dimensional imaging.

Airborne and spaceborne platforms for remote sensing

Remote sensing normally assumes that information is collected from a distance. The distance can range from hundreds of feet to thousands of miles. While aerial photographs were taken from hot-air balloons in the early history of remote sensing, the airplane has provided the first fully controllable aerial platform for remote sensing of the Earth since the beginning of the twentieth century. Modern aircraft, both helicopters and fixed wing, provide multiple vantage points. Detailed images are acquired from hundreds of feet above ground by helicopters, and most aerial photographs are obtained from fixed-wing aircraft at altitudes between 1,000 and 40,000 feet above ground level.

A recent addition to the aerial platforms available for remote sensing are pilotless aircraft called *unmanned aerial vehicles* (UAV) or, increasingly, *unmanned aerial systems* (UAS), because they include ground-based navigation software and various imaging devices. These platforms are usually much less expensive to operate per hour than piloted aircraft. They are ideal for small area surveys because they have limited range and altitudes. Several small (wingspans less than 5 feet)

UASs are available from commercial companies that have shown utility for detailed aerial imaging (Trimble 2015). Small helicopters are also popular. Large UASs (wingspans greater than 5 feet) can provide greater range and flight time than small ones. Early adopters of UAS in the United States are prevented from commercial operations by the Federal Aviation Administration, which is currently (2015) developing rules for flying these systems in US airspace. Law enforcement agencies and some universities are currently operating UASs in the United States on a not-for-profit basis. Tens to hundreds of detailed images can be acquired by UASs on a single flight, and these are pieced together into a map-accurate, mosaic image, as described in chapter 6.

Numerous orbiting satellites serve as space-based platforms for remote sensing. The most common type of orbit is a near-polar, low-Earth orbit (LEO; 200–500 miles altitude) that circles the Earth approximately every 100 minutes, from pole to pole. The orbital period is set by the velocity required to resist Earth's gravity (about 17,000 miles per hour) and the altitude of the orbit. A useful geometry is obtained because the orbit is near polar with the Earth spinning underneath, west to east, and the satellite orbiting north to south. The Earth has rotated to the east during the 100 minutes that it takes the satellite to return to the same location north-south, so each successive orbit track is west of the previous one (figure 2-8).

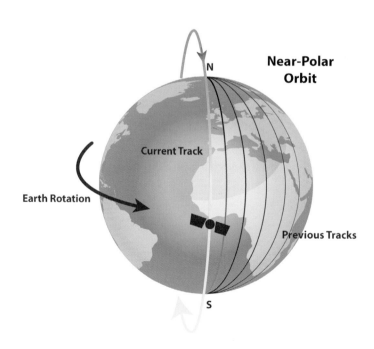

Figure 2-8. A diagram showing the relation between a polar orbit and the sequential ground tracks imaged underneath the orbit.

The local time is near equal for both orbit tracks because 100 minutes have passed and the Earth has rotated about 1.5 time zones. Multiple images can be stitched together having approximately the same sun angles and resulting shadow patterns among them because of this creative orbit design. However, the gaps between successive orbit tracks must be filled in with imagery acquired over a period of approximately 1 to 15 days, allowing variable cloud patterns and atmospheric conditions to change the brightness qualities of adjacent images. Multiple satellites in near-polar orbits can be used to acquire adjacent images over shorter time periods. The revisit time is also reduced when pointable optics are used to look forward and back or side-to-side to acquire imagery of the same spot during the same orbit or on successive orbits of a single satellite. These systems allow operators to decrease the time between image acquisitions to as little as several minutes. The same local sun angle is not achieved when the revisit time is on the order of hours.

A variation of the polar orbit, the inclined orbit is used for human space flight in the International Space Station, which keeps the satellite directed over the more populated parts of Earth between about 50 degrees north and south latitude. Only opportunistic Earth imaging is done on a sporadic basis from crewed space vehicles in inclined orbits.

Another type of creative orbit geometry is used for telecommunications satellites and some meteorological satellites: the geostationary orbit. This is an equatorial orbit with the altitude adjusted so that the satellite completes one orbit in 24 hours, synchronized to match the rotation of the Earth. Thus, the satellite appears to park above the equator at a specific longitude (figure 2-9).

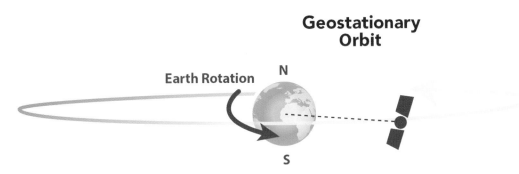

Figure 2-9. An illustration of the geostationary orbit.

The geostationary orbit is ideal for obtaining images acquired in rapid succession of the same location on the Earth. The problem with this orbit is that the altitude must be about 22,000 miles in order to synchronize the required velocity of the satellite with Earth's rotation. The altitude of geostationary satellites is about 50 times higher than low-Earth orbit, precluding the acquisition of highly detailed imagery. Geostationary meteorological satellites do not require fine detail because weather fronts are hundreds of miles across and can be imaged nicely with 1- to 4-kilometer pixel widths. Geostationary orbits are used to provide images of a full Earth disc on a 30-minute repeat cycle, which can be shortened depending on monitoring requirements for specific weather events. In addition to meteorological applications, the images can be used for regional mapping of general land use and land cover, in cases where fine detail is not desired.

Remote sensing review: sensor types, methods, wavelengths, and sources

Table 2-1 provides a summary of the relationship between the sensor type, method of remote sensing, wavelengths of EM radiation used, and the sources of that EM radiation.

Table 2-1. General methods of remote sensing.

Sensor type	Method of remote sensing	Wavelengths used (micrometers)	Source of radiation
Optical	Passive—Reflected radiation	Short (0.4–3.0)	Sun
Thermal	Passive—Emitted radiation	Long (3.0–14.0)	Earth
Laser–lidar	Active—Reflected radiation	Short (0.5–2.0)	Sensor
Microwave–radar	Active—Reflected radiation	Very long (10,000–300,000)	Sensor

Optical instruments passively measure reflected short-wave EM radiation (light) generated by the sun. Thermal sensors passively measure long-wave EM radiation (heat) emitted from the Earth's surface that is related to the temperature of the surface. Lidar systems generate short-wavelength radiation actively to produce three-dimensional point clouds from the reflected returns. Imaging radar generates very long wavelength microwaves actively to create unique images from the reflected returns in all weather conditions.

Small amounts of microwave EM radiation are also emitted from the Earth, and these can be sensed passively. Specialized sensors are required to capture this weak microwave radiation, and the software required to process these images is beyond the capabilities of GIS (Lillesand, Kiefer, and Chipman 2008).

Chapter 3
Effects of the atmosphere on image quality

When working with images, you must understand the effects of the atmosphere on image quality. In this chapter, you will learn how the atmosphere and cloud cover affect the electromagnetic (EM) radiation detected by remote sensing systems. You will also learn how remote sensing systems work to minimize atmospheric interference.

Almost all remote sensing systems use EM radiation to image Earth features. Before that radiation reaches the sensor, it must pass through the atmosphere first. The atmosphere and cloud cover can interfere with EM radiation by absorbing or scattering it. Thus the atmosphere can affect the brightness of remote sensing images, as does the amount of EM radiation reflected or emitted from the Earth's surface.

When mapping surface features with imaging GIS, the reflected or emitted radiation is the *signal*, and the atmospheric contribution is the *noise*. Remote sensing instruments are normally programmed to focus on wavelengths that are less affected by the atmosphere. Atmospheric effects are therefore minimized and do not normally change the relative brightness values of surfaces enough to preclude accurate image interpretation. However, under some conditions the atmosphere can affect brightness enough to cause confusion. Fortunately for users of remote sensing, the signal is usually stronger than the noise.

When sensing from Earth orbit (satellite remote sensing) using solar radiation, the radiation must pass through the entire atmosphere twice—from the sun to the Earth's surface and back up to the orbiting satellite (figure 2-1). When sensing from airplanes, less atmosphere must be traversed from the Earth to the sensor. However, the densest part of the atmosphere lies below 20,000 feet of altitude, and most images from airplane platforms are acquired from between 3,000 and 20,000 feet above the ground. Aircraft flying at 10,000 feet above ground experience almost as much atmospheric effect as aircraft at 40,000 feet. All aerial vantage points incur some form of atmospheric effect.

GIS software can be used to correct for atmospheric effects. However, the best methods of correcting for atmospheric effects require specialized image processing software. This software is described briefly in chapter 7 with suggested further reading. Image providers often correct for atmospheric effects as a service to their customers. Understanding these corrections and how they affect the image is necessary for accurate interpretation when working with corrected imagery, as explained in chapter 7.

The atmosphere affects the amount of EM radiation received at the sensor in two ways: (1) by absorbing some of the radiation to darken the image and (2) by scattering the radiation to lighten the image. These counteracting effects do not cancel each other out because they occur at different wavelengths.

Radiation transmission through the atmosphere

The atmosphere is transparent only in some wavelength regions of the EM spectrum. People generally think that the atmosphere is transparent because human vision occurs in a transparent wavelength region of the EM spectrum. The atmosphere is actually opaque to some wavelength regions. Therefore, sensor designers filter instruments to avoid opaque regions. The transparency of the atmosphere depends on the wavelength of the EM radiation moving through it (figure 3-1).

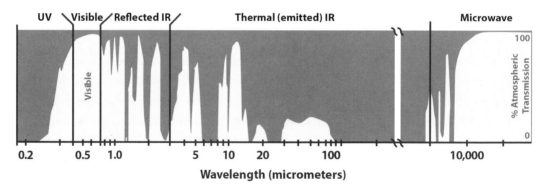

Figure 3-1. Atmospheric transmission of electromagnetic radiation in the wavelength regions commonly used for Earth imaging.

Modified from NASA Earth Observatory image at http://earthobservatory.nasa.gov/Features/Remote Sensing/remote_04.php.

The atmosphere is essentially transparent in the visible range between 0.4 and 0.7 micrometers. It is nearly transparent in the near infrared, up to 1.3 micrometers. Wavelength regions with maximum atmospheric transparency are called *atmospheric windows* (shown in light gray in figure 3-1). Digital aerial photography, the most common type of remote sensing, operates in the large atmospheric window between 0.4 and 1.3 micrometers.

Ozone absorbs most of the ultraviolet (UV) spectrum at wavelengths below 0.4 micrometers, precluding remote sensing in the UV wavelength region unless it is done at wavelengths that are close to the visible range (that is, close to, but slightly less than 0.4 micrometers). Other gases, including water vapor, absorb strongly in three regions centered at approximately 1.4, 1.9, and 2.8 micrometers. Remote sensing must be done carefully in this short-wave infrared (SWIR) spectral range between 1.3 and 3 micrometers because of variable water vapor in the atmosphere that can shrink the width of these windows. Sensor designers reduce the effects of the strong water absorption bands by centering spectral bands at about 1.6 and 2.2 micrometers, in the windows between opaque regions. SWIR bands are becoming increasingly popular for remote sensing because of the unique reflectance properties of Earth surfaces in these wavelength bands even though atmospheric attenuation can darken the image when sensing under conditions of high atmospheric water vapor.

Thermal sensing of EM radiation emitted from the Earth is commonly done between 8 and 12 micrometers and, more rarely, between 3 and 5 micrometers to take advantage of the atmospheric windows present at those wavelengths. The wavelength of maximum emission for an ambient temperature Earth surface is 10 micrometers (figure 2-2). This is centered in the atmospheric window in the range of 8–12 micrometers (an advantageous alignment for remote sensing). Very hot objects such as flowing lava and wildfires (approximately 800°C) have a shorter wavelength of maximum emission than ambient temperature surfaces. The thermal window between 3 and 5 micrometers is often used for sensing these phenomena.

Strong absorption characterizes the atmosphere in the infrared region beyond 12 micrometers, precluding remote sensing at these wavelengths. However, there is virtually no solar radiation and very small amounts of Earth-emitted radiation in this wavelength range, making sensing very difficult regardless. Finally, the atmosphere becomes completely transparent at wavelengths longer than 10,000 micrometers (1 centimeter), which is the region where microwave radar systems operate.

Scattering of radiation by the atmosphere

The atmosphere scatters radiation in addition to restricting transmission at certain wavelengths. This scattered radiation finds its way into the sensor and adds extra brightness to the image. This extra brightness is not correlated with features on the Earth's surface. The extra brightness reduces reflectance differences of surface features, thereby reducing contrast in the resulting image (figure 3-2).

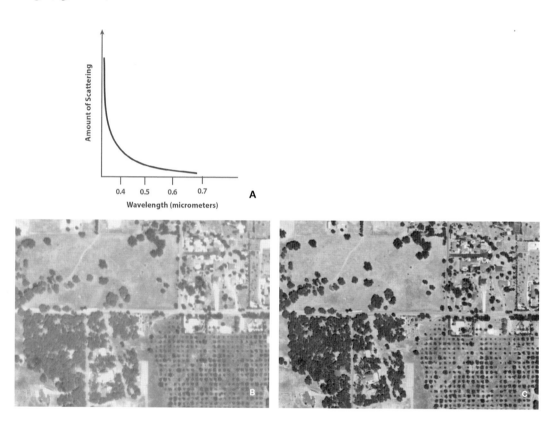

Figure 3-2. Atmospheric scattering of short-wave electromagnetic radiation.

Courtesy of National Agriculture Imagery Program, US Department of Agriculture, Farm Service Agency, California Department of Fish and Wildlife, Biogeographic Data Branch.

Figure 3-2A shows the inverse relationship between wavelength and the amount of scattering for solar radiation passing through the atmosphere. Scattering increases exponentially as wavelength becomes shorter. Humans perceive the sky as blue because short, visible wavelengths of sunlight are scattered by the atmosphere, and humans perceive these wavelengths as blue light. Below the graph are portions of an aerial photograph of oak trees and an orchard with a dead grass understory. Figure 3-2B is an image from the short-wave end of the visible spectrum centered at 4.5 micrometers. Figure 3-2C is an image from the long-wave end of the visible spectrum centered at 6.5 micrometers. Note the increase in image brightness and the decrease in image contrast shown in B compared to C because of the increased atmospheric scattering at shorter wavelengths.

Aerial photographers normally filter out the short end of the visible wavelength region between 0.4 and 0.5 micrometers to reduce brightness from scattering when creating black-and-white images or false-color images that avoid these blue wavelengths. This minus-blue filtering can upset color balance for natural-color photography because the blue portion of the visible spectrum (0.4–0.5 micrometers) is filtered out of the sensor. A satisfactory compromise for natural-color photography is achieved by filtering out only part of the first third of the visible spectrum (that is, wavelengths below 0.45 micrometers). More information on color, including a complete discussion of additive color theory and false-color infrared imagery, is provided in chapter 5.

Atmospheric scattering can also be reduced by avoiding early morning or late afternoon hours. Solar radiation must travel through a greater thickness of atmosphere when the sun is low in the sky because of the angled path. Not only is the haze effect reduced, but the intensity of the solar radiation is greater near solar noon. Shadows are reduced in length, making the middle of the day an ideal period for Earth imaging.

Not only do the gases in the atmosphere scatter short-wave EM radiation, but small particles that are often suspended in the atmosphere also scatter radiation. Scattering from particles can affect the entire solar spectrum. Air pollution, haze, and smoke can scatter so much radiation that remote sensing of the Earth becomes impossible at wavelengths below 3 micrometers. Thermal infrared wavelengths longer than 3 micrometers are much less affected by suspended particles, and long-wave thermal sensors are often able to penetrate smoke. This provides a double benefit for thermal infrared remote sensing for wildfire imaging: the thermal wavelengths are not scattered by the smoke, and the brightness of the image indicates hot spots in the fire (figure 3-3).

The image shown in figure 3-3A was acquired in the short-wavelength spectrum between 0.4 and 1.0 micrometers of an area near Bend, Oregon, while a wildfire was burning. Note the excessive atmospheric scattering from smoke (center right) and high cirrus clouds (top third) that obscure the scene below. The image in figure 3-3B is the same area captured at the same instant in the thermal infrared wavelength region. Note the dark areas of the clouds and snow fields, which are colder than the surrounding landscape, and the bright areas (center right), which are the actively burning areas of the wildfire.

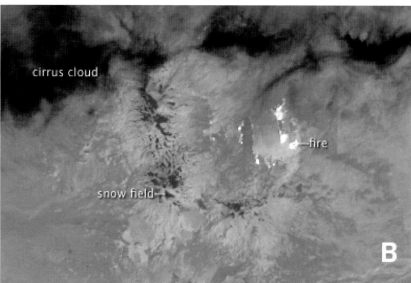

Figure 3-3. An example of two images of a wildfire acquired simultaneously with the same multispectral sensor. *A*, Smoke dominates the image from the visible spectrum. *B*, Fire hot spots are shown in the thermal infrared image.

ASTER (Advanced Spaceborne Thermal Emission and Reflection Radiometer) satellite imagery, near Bend, Oregon, September 19, 2012, courtesy of NASA.

Ground-obscuring clouds

Clouds are composed of water droplets that scatter and absorb EM radiation entering them so much that they become opaque to short-wavelength solar radiation and long-wavelength thermal radiation. When flying above the clouds, airborne sensors, operating between 0.4 and 12 micrometers, acquire images of clouds rather than the landscape below. Aircraft flights are therefore normally scheduled for near-cloud-free days. Satellites acquire imagery at set time intervals based on their predictable orbit period, which cannot be altered to avoid cloud cover. However, seasonal cloud-free satellite imagery (that is, four images per year) is usually available as long as imagery comes from multiple satellite sensors. Cloud cover is especially limiting to Earth imaging in the tropics, where some geographic regions have virtually continuous cloud cover throughout the year. This cloud cover precludes both satellite- and aircraft-based remote sensing at wavelengths below 12 micrometers.

Clouds are not always obscuring barriers to remote sensing. Images of clouds are desired for meteorological applications, and National Oceanic and Atmospheric Administration (NOAA) weather satellites commonly provide imagery of clouds from the visible and infrared spectra. The temperature of the clouds is correlated with precipitation (colder clouds tend to be associated with more precipitation), and therefore thermal infrared images are commonly collected by weather satellites. Users of thermal infrared weather satellite imagery will notice that the tones are reversed from normal convention by NOAA. Clouds, which are usually cooler than water or land surfaces, are light toned rather than dark toned on the imagery.

Images taken from under cloud cover can generate essentially shadowless conditions that provide more detail in shadows from buildings or trees than imagery taken from a clear sky. However, there is seldom enough cloud cover to provide near-constant diffuse lighting for this to work.

The only way to remotely sense through clouds is to use very long wavelength "microwave" imaging. The vast majority of microwave sensing is done actively with imaging radar. This active sensor uses wavelengths of 10,000 micrometers (1 centimeter) and longer that effectively penetrate cloud cover and all haze and smoke. Weather radar commonly uses a wavelength of 1–8 centimeters that penetrates clouds but not precipitation, thus producing a return signal and therefore an image of precipitation. The radar return signal, when combined with a thermal image of the cloud cover, can indicate what portions of the cloud cover are producing precipitation. Very long wavelength (approximately 20-centimeter) imaging radar penetrates even the heaviest precipitation as well as all cloud cover. These active sensors are used to image surface features through the most severe weather conditions.

Chapter 4
Creating two-dimensional images with sensors

To effectively use the two-dimensional images generated by remote sensors, you need to understand how the sensors work. In this chapter, you will learn how remote sensing instrumentation influences the attributes of the images it produces rather than the engineering details of the sensors. You will learn about the instruments that generate two-dimensional images, including cameras, multispectral sensors, and imaging radar. Instruments that generate three-dimensional elevation models and point clouds, such as specialized interferometric radars and laser scanning (lidar), are described in chapter 6. This chapter concludes with a summary of the two-dimensional imagery types generally available for input to GIS.

Passive sensors

Most passive sensors are called *optical sensors* because they focus electromagnetic (EM) radiation using a system of lenses and/or mirrors to generate images with either solar- or Earth-generated radiation. The simplest sensor employed to capture short-wave radiation passively is the camera. The image is formed by focusing reflected solar radiation onto a two-dimensional "image plane" through a lens (figure 4-1).

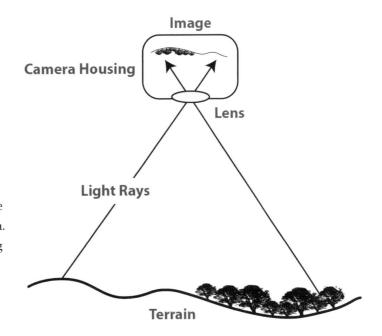

Figure 4-1. A diagram of a simple camera.

The simple lens in the diagram is actually a complex system of lenses in aerial mapping cameras that focus a nearly distortion-free image. The image is recorded by a matrix of very small, light-sensitive detectors that produce an electrical signal when impacted by EM radiation (Campbell and Wynne 2011). That signal is digitized by the camera's electronic system into numbers that represent the brightness levels of a digital image. Each detector produces a brightness value for one picture element (pixel) in the resulting image (chapter 1). The position of the detectors on the image plane determine the x,y location of the pixels. Before the current millennium, the imaging plane of the camera was commonly overlaid with photographic film that recorded a chemical response to the incoming radiation. Although film is still used occasionally for Earth imaging today, it is normally scanned into digital values to form a digital image before delivery to GIS professionals.

This simple idea of one lens and one image plane is complicated by the desire to create sensors that record multiple images in multiple wavelength bands of the EM spectrum. The simplest mass-market digital cameras divide the visible spectrum into three wavelength bands to create color photographs. More information on color, including color vision theory, is presented in chapter 5. Color cameras contain patterns of three or four detectors for each pixel on the image plane to record incoming radiation in multiple wavelength bands. Four bands are often acquired to allow the creation of images with sensitivity in the near-infrared wavelength spectrum, in addition to the visible range (chapter 2).

Modern aerial cameras are much more complicated than the diagram shown in figure 4-1. In modern aerial cameras, multiple image planes are often combined to generate a large-sized multispectral image, which increases their spectral sensitivity, image detail, and field of view. The diagram is useful for understanding the point perspective geometry of cameras, which creates displacements in images that are caused by variation in terrain elevation, which is further explained in chapter 6.

Remote sensing using wavelengths longer than about 1.3 micrometers presents even more technical challenges to sensor designers than sensing with short wavelengths. Complicated multispectral sensors employ multiple sensor arrays in complex geometries (Lillesand, Kiefer, and Chipman 2008). Recording images from wavelengths that are longer than 1.3 micrometers requires sophisticated optical and electronic design. Sensors become even more complex when sensing thermal radiation at wavelengths beyond 3 micrometers, often requiring cryogenic cooling of detectors. These complex sensors employ mirrored telescopes to achieve fine detail from high altitudes, including Earth orbit.

Active sensors

Active remote sensing is primarily employed today in two wavelength regions: short-wave light detection and ranging (lidar) in the visible and near-infrared bands and very long-wave radio detection and ranging (radar) in the microwave band (chapter 2). Lidar systems produce three-dimensional (3D) point clouds rather than two-dimensional (2D) images and are therefore explained in chapter 6. Radar systems are primarily used to generate two-dimensional images and are described here. Radar can also be used to generate a three-dimensional surface model using an interferometric approach, as described in chapter 6.

Radar is the only remote sensing technology that penetrates clouds and precipitation because of the very long wavelength EM radiation employed; it is a truly all-weather sensor (Richards, Scheer, and Holm 2010). Very long wavelength (20- to 30-centimeter) radars can even penetrate dry sand dunes, revealing ancient buried river channels. Radar is especially useful for sensing in extremely cloudy regions near the equator and the Earth's poles. As a result, several national remote sensing programs include a satellite-mounted radar sensor: the European Space Agency, Canadian Space Agency, and Japan Aerospace Exploration Agency are examples. These systems produce all-weather radar images with fine image detail (5 meters or better). The common product produced from these systems is a two-dimensional image (figure 2-7). Radar sensors actively generate a pulse of micro-wave radiation that propagates toward the Earth's surface in a side-looking geometry (figure 4-2).

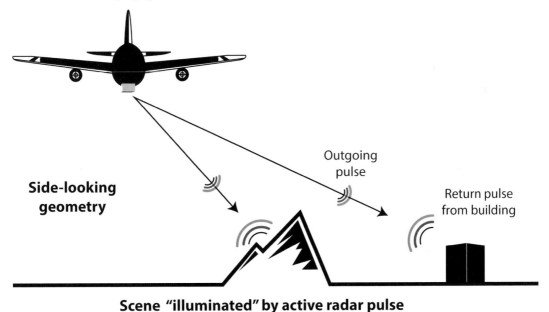

Figure 4-2. The side-looking geometry of the energy flow profile for an imaging radar system.

Radar sensors actively create an image by precisely measuring the time required for the reflected pulse of microwave radiation to return to the sensor. The radar antenna is both the source of the radiation and the sensor for recording the returning pulse. The time can be accurately converted to distance for the round trip due to the constant velocity of EM radiation. If the pulse were directed vertically downward, the sensor could calculate the flying height of the sensor above ground, dividing the time required for the round trip by two. This, in fact, is done to manufacture a radar altimeter, a nonimaging device.

Imaging radar, however, requires a side-looking geometry, so the horizontal distance traveled by the return pulse can be calculated from that time required for the return to reach the sensor. Returning reflected pulses arrive at different times, with those traveling from greater distances taking longer to return. Thus, the image is assembled from these returns, with returns from the near range arriving sooner than returns from the far range of the resulting image. The sensor illuminates a narrow swath of the image area in the very brief period required for the reflected pulses to return. The platform moves forward (coming straight out of the page in figure 4-2), and then the next adjacent swath is illuminated. Both the wavelength and polarization of the microwave radiation is controlled and independently sensed, with modern radar systems adding to the information content of this all-weather imagery (Richards, Scheer, and Holm 2010). Processing original radar return signals into images is not yet possible with GIS software, but display and enhancement of radar multipolarization and multiwavelength radar imagery is accomplished in GIS as it is with any type of multispectral imagery. Radar, however, records very different reflectance properties of the landscape compared to optical images because of the extremely long wavelengths involved. Further reading is recommended for the most effective interpretation (Richards, Scheer, and Holm 2010).

General image attributes: The four Rs

An understanding of four attributes of images (often called the *four Rs*) allows practitioners to evaluate different types of remote sensing images for many applications regardless of the sensor technology used to produce them. The four Rs are four different characteristics of resolution, which is the process or capability of making distinguishable the individual parts of an object. These characteristics are *spatial*, *spectral*, *radiometric*, and *temporal*.

Spatial resolution

Spatial resolution is by far the most common meaning of resolution inside and outside the field of remote sensing. In fact, most people equate the term *resolution* with spatial resolution. The size of the pixel (chapter 1) on the ground (the ground sample distance, GSD) determines the amount of spatial detail visible in the resulting digital image. The size of the pixel does not indicate the spatial resolution of the image, however. Earth features must be composed of several pixels in order to recognize shape and size, and these attributes are often the key to identifying them. For example, a limited-access expressway is clearly visible running vertically near the center of the 1-meter spatial resolution image in figure 4-3B, and a faint linear feature representing this same road is barely visible in the 25-meter spatial resolution image of the same area in figure 4-3A. Without the extra detail of the interchange, for example, an interpreter has a hard time identifying the linear feature as a limited-access highway.

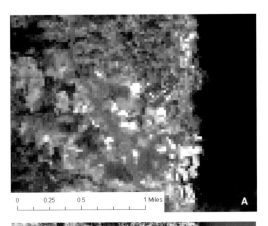

Figure 4-3. A comparison of features visible at 25-meter spatial resolution (*A*) with features visible at 1-meter spatial resolution (*B*). Progressive enlargements of the 1-meter resolution image are presented in *C* and *D*.

Courtesy of US Geological Survey: (A) is a portion of a 1996 Landsat satellite image; (B), (C), and (D) are portions of 1993 digital orthophoto quarter quads (DOQQ), Clear Lake, California.

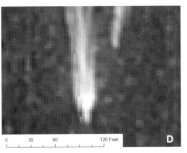

Of course, the 1-meter resolution image contains more spatial detail than the 25-meter resolution image, but remember that spatial resolution is always limited at some point. Zooming in to the 1-meter image somewhat (figure 4-3C) makes visible smaller objects that are unnoticed at the scale shown in figure 4-3B. Further zooming until the pixels are visible (figure 4-3D) reveals the limitation of enlargement to render objects that are distinguishable from the background.

For example, when the v-shaped feature on the lake in the right-hand portion of the 1-meter spatial resolution image (figure 4-3B) is enlarged 10 times (figure 4-3C), two v-shapes become visible. Those familiar with recreational boating in California will interpret the v-shapes as wakes from a speed boat towing a water skier (or wake boarder). Those more familiar with marine warfare with no information as to the location of the image may interpret the shapes as something very different. So, experience and context are important. Note that the skier is not visible, and neither are the details of the boat when the image is further enlarged (figure 4-3D). Clearly an object must be composed of many pixels before it is positively identifiable. The dimension of the pixel determines the lower limit of spatial detail for an image but does not indicate positive identification of objects of that size. However, even a few pixels when viewed in the correct context can be identified with some confidence (as a speed boat and water skier in this case) without information on specific shape or details of the boat or the skier.

Spectral resolution

Spectral resolution refers to the number and width of the spectral bands available in the image. A spectral band represents a range of wavelengths to which the sensor is sensitive. For example, black-and-white aerial photography is normally sensitive to only one spectral band that spans almost the entire visible spectrum (0.5–0.7 micrometers). Thus, black-and-white aerial photography is commonly referred to as *panchromatic* (sensitive to all colors, rendered in shades of gray). Color photography expands spectral resolution to three bands representing the three primary additive colors (chapter 5). Image information is greatly increased when a fourth spectral band is added in the near infrared. A comparison of an image generated from a visible spectral band and an invisible-made-visible, near-infrared spectral band is shown in figure 4-4.

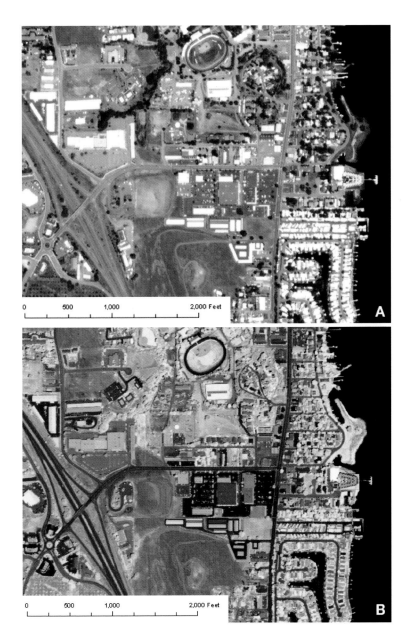

Figure 4-4. West shore of Clear Lake, California, summer 2009. A portion of an aerial photograph from a visible wavelength band (0.6–0.7 micrometers) is shown in A. The same area from a near-infrared band (0.7–0.9 micrometers) is shown in B.

Courtesy of National Agriculture Imagery Program, US Department of Agriculture, Farm Service Agency, California Department of Fish and Wildlife, Biogeographic Data Branch.

Near-infrared radiation reflects very differently than visible radiation. Water and paved roads are darker and more discernible from other features in the near-infrared band. Trees are darker and more discernible from nonvegetated surfaces in the visible band.

Central to the idea of spectral resolution is the goal of capturing the spectral signature (chapter 2). Figure 4-5 displays a graph of spectral reflectance signatures for common Earth surfaces and includes the approximate width and location of spectral bands common to many sensors.

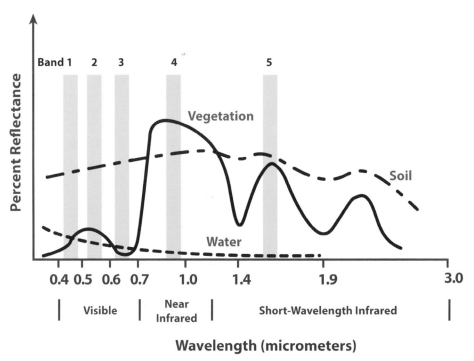

Figure 4-5. The position of five spectral bands common to many sensors overlaid on spectral reflectance signatures of three common Earth surfaces.

Bands 1 through 3 describe the visible spectrum and are necessary for forming natural-color images. These bands are common to most digital cameras. Band 4 in the near-infrared spectral region always shows water as very dark and vegetation as bright. It is critical for vegetation analysis and enhances water bodies by rendering them in dark shades. Moreover, different types of vegetation are displayed in various bright shades, allowing for enhanced vegetation-type mapping in the near-infrared band. These four bands are included on most digital aerial cameras and most Earth resources satellites. The fifth band in the short-wavelength infrared (SWIR) is included on some Earth resources satellites and is useful for determining moisture content of various surfaces. Reflectance signatures in the SWIR band are similar to those of the visible bands, yet the SWIR band usually provides superior contrast compared to the visible spectrum bands, especially between vegetation and water.

One could argue that spectral resolution is adequate once the spectral signature is described by the spectral bands, that is, when a sufficiently narrow spectral band is placed between or at each inflection point of the spectral signatures of common Earth surfaces. This is essentially achieved with five spectral bands of approximately 0.05 micrometers in width, as shown in figure 4-5. That said, more and narrower bands generally do provide more spectral information about the surfaces being imaged. However, as the band width is reduced, the amount of radiation caught by the sensor decreases, making it difficult and expensive for sensor designers to create high-quality images when using very narrow bands. The three visible bands with the addition of a fourth near-infrared band provide excellent spectral detail for many applications, and this degree of spectral resolution is common in many sensors.

Sensor choices become limited for applications requiring more than four bands. However, many satellite sensors have 5 to 11 bands, including a thermal infrared band. Images from these sensors are commonly available from various national space-based remote sensing programs (e.g., the US Landsat program, the Indian Remote Sensing [IRS] program, and the French SPOT satellite system; see the table at the end of this chapter). The added spectral information from these satellite sensors somewhat makes up for their lack of spatial resolution (1.5−30 meters) from satellite orbital altitudes. However, since the turn of the millennium, commercial satellite companies have greatly improved the spatial resolution of satellite remote sensing. They provide satellite images commercially with 0.3-meter spatial resolution having four spectral bands: three visible and one near infrared (details later in this chapter).

Engineers have introduced sensors with approximately 200 very narrow spectral bands in the last 20 years. Because the number of bands is so much greater than ever available before, these systems are called *hyperspectral* instead of multispectral. These instruments represent very fine spectral resolution and essentially provide imaging spectroscopy from the air. Specialized software is necessary to manage hundreds of bands, but the information obtained from these systems can be impressive depending on the application. Specific chemical salts have been identified on desert dry lakes from the air, for example. Military applications include finding camouflaged vehicles under tree canopy that are not detected with traditional multispectral imaging. Eventually GIS software may include hyperspectral image analysis functionality, but for the near term this advanced imagery will be analyzed by remote sensing specialists (Campbell and Wynne 2011).

Radiometric resolution

Radiometric resolution describes the number of brightness levels present in the image, called the *quantization* of the image. More levels produce more information. Humans can discern about 30 shades of gray (with practice) on step wedges of progressively brighter gray chips. Computer words (bytes) are the basis of almost all computer software. Bytes contain 8 bits of binary data and are commonly used to store letters and numbers in computer code. Thus, the byte is a convenient packet in which to store the brightness value of a pixel. Because an 8-bit byte stores binary number values between 0 and 255 (that is, 2 to the 8th power, producing 256 values, including 0) and 256 brightness values exceed human discernment, the byte became the standard for image quantization (figure 4-6).

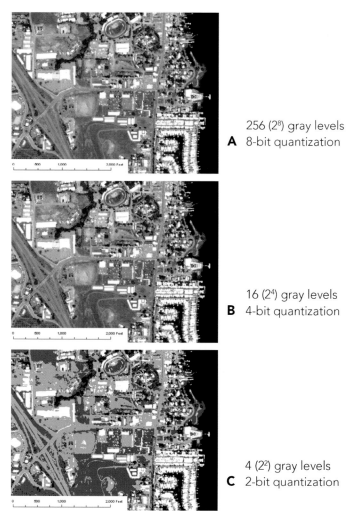

256 (2^8) gray levels
A 8-bit quantization

16 (2^4) gray levels
B 4-bit quantization

4 (2^2) gray levels
C 2-bit quantization

Figure 4-6. A comparison of three quantizations for a single-band image from the visible spectrum. The standard 256-level quantization is shown in image *A*, 16 levels in image *B*, and 4 levels in image *C*.

Image courtesy of US Geological Survey. Image modifications by author.

The radiometric quality of an image does not appear to degrade significantly until the number of gray levels is greatly reduced to four (figure 4-5C). However, sensor improvements since the turn of the millennium have made it possible to record slight differences in radiation over much greater ranges of brightness than was available with earlier technology. Modern digital sensors are showing details more effectively in the shadows and in the very bright regions of an imaged area than film cameras. As a result, imaging sensors now record digital images with more gray levels than 8-bit quantization, including 11-bit quantization having 2,048 levels. These 11-bit image data are commonly packed into two computer bytes per pixel. This forms a 16-bit image file format having 65,536 levels but only using the first 2,048. Of course, these are many more levels than human vision can discern. However, the GIS analyst is able to assign a range of visibly discernible gray levels (or color levels) to various ranges of brightness, thereby enhancing either shadows or bright areas (explained in more detail in chapter 5). Automated image classification procedures are also able to exploit the large range of quantization available in modern imagery, as further explained in chapter 8.

Temporal resolution

Temporal resolution refers to the time elapsed between image data collections. The majority of commercial remote sensing from aircraft and satellites produce still images, and the revisit schedule of the aircraft or satellite determines the rapidity of image coverage: yearly, monthly, etc. The timeliness requirement of the imagery is dependent on the application. Images obtained at the height of a flood can help determine which areas are flooded, for example. Subsequent imagery acquired at specified time intervals after the high water can document receding flood levels. The timing of overflights is dependent on the dynamics of the event being monitored. Airplanes provide great flexibility in timing of image collection, which allows analysts to capture the dynamics of the event and avoid ground-obscuring clouds.

Many Earth imaging satellites providing spatial resolution of 30 meters or finer have repeat cycles of about 15 days, depending on orbit parameters. Some modern imaging satellites have side-to-side pointable optics that greatly reduce the time interval for repeat coverage of specific locations, down to one orbit of about 100 minutes. Even shorter repeat cycles are possible with fore-and-aft pointing optics. Also, because of the large number of imaging satellites currently in orbit, images acquired from multiple satellites can greatly increase the temporal resolution of images (of similar type, from multiple satellites), down to daily coverage or even multiple images per day.

Imaging engineers have also created motion images (digital video) to provide the finest temporal resolution available by remote sensing. The most familiar form is full-motion video, which obtains 30 to 60 images per second. Full-motion video is used for military and security applications in which moving features need to be tracked in a continuous monitoring situation.

Data storage and transmission requirements are tremendous for full-motion video and, as a result, spatial resolution and spectral resolution is often reduced to decrease the data load. The field of view of these systems is also reduced to decrease data loads. A less data-intensive form of motion imagery called *wide area motion* is used in security applications. These images are obtained by airborne systems that acquire one to two images per second, thereby greatly reducing the data loads compared to full-motion video. They can cover a large field of view compared to full motion because of these reduced data loads. These systems are used to track moving objects in near real time. Most GIS software does not display georeferenced video imagery in a map viewer but can display video with third-party media viewers activated by a mouse click on a point. Video functionality should increase as motion imagery becomes more available in the future. Skybox Imaging has recently put satellites in orbit that capture video (Skybox Imaging 2014). Skybox has also recently (2014) signed an agreement to be acquired by Google.

Common sources for remote sensing imagery

Two comprehensive references for satellite imagery sources and types are available electronically from the American Society for Photogrammetry and Remote Sensing on their website (Stoney 2008; Williams and Brown 2006). The following paragraphs describe the most common sources of imagery available in the United States and throughout the world. Government acronyms, key words, and corporate names are listed so they can be used for searches on the Internet. Information is also summarized by a table at the end of this section.

You can acquire images from satellites managed by the US government for free. To acquire images from commercial satellites or from other governments, you may have to purchase a license. Prices and sources change often. For current pricing, contact the organizations that provide the images you are interested in using.

US weather satellites are well-known sources of free satellite imagery, available over the Internet. Spatial resolution is poor, varying from 1–10 kilometers. The 10-kilometer images are most often used for atmospheric and ocean analysis. The 1-kilometer images can be useful for land applications, including surface/hydrology, snow/ice, and vegetation mapping on a regional to continental scale. Images are available worldwide from the National Oceanic and Atmospheric Administration (NOAA), National Environmental Satellite, Data, and Information Service (NESDIS): http://www.nesdis.noaa.gov.

Landsat, the Earth satellite system created by the National Aeronautics and Space Administration (NASA) and maintained by the US Geological Survey, provides the most well-known Earth resources satellite images. These images are useful for numerous regional applications in a GIS. The Landsat program has been in existence since 1972 and has launched eight satellites over

40 years (Lillesand, Kiefer, and Chipman 2008). Spatial resolution is moderate (15- to 30-meter pixels). Spectral resolution is rich, with three bands in the visible spectrum, one in the near infrared, two bands in the short-wavelength infrared, and one in the thermal infrared. Enhanced sensors in later versions of Landsat add one fine spatial resolution, panchromatic band sensitive to visible and near-infrared EM radiation. The spatial resolution of multispectral Landsat imagery is 30 meters, and the panchromatic imagery is 15 meters. Computer algorithms are commonly used to merge the resolutions, producing a 15-meter resolution, multispectral image, as explained in chapter 7.

While sensor malfunctions degraded Landsat imagery from Landsats 5 and 7 in the last few years, a new Landsat satellite with improved spectral resolution (Landsat 8) was launched successfully in the spring of 2013. This satellite (the Landsat Data Continuity Mission, LDCM) provides an additional visible band in the blue spectrum, an additional SWIR band for cirrus cloud detection, and two thermal infrared bands rather than the one thermal band of previous Landsat sensors. Landsat images are available from three government websites:

- EarthExplorer: http://earthexplorer.usgs.gov
- Global Visualization Viewer: http://glovis.usgs.gov
- LandsatLook Viewer: http://landsatlook.usgs.gov

Landsat imagery is also available from a few university websites (e.g., University of Maryland, Global Land Cover Facility: http://www.landcover.org).

In addition, GIS software can be connected to Landsat data servers to provide Landsat imagery interactively over the Internet to the GIS (e.g., Esri Landsat image services: http://www.esri.com/software/landsat-imagery).

The French satellite remote sensing program, SPOT, has been in existence since 1986 and is also well known. The French system boasts finer spatial resolution than Landsat (1.5-meter vs. 15-meter pixels) but poorer spectral resolution (5 spectral bands vs. 11). SPOT is a European Union, commercial–government program that has recently (2014) sold its SPOT 7 satellite to Azerbaijan. SPOT imagery continues (spring 2015) to be marketed through Airbus Defence and Space: http://www.geo-airbusds.com.

The spatial resolution of satellite images was improved tremendously at the turn of the millennium with the introduction of two commercial Earth resources satellites: IKONOS (1999) and QuickBird (2001). The two companies that launched these and the World View satellite in 2007 have since merged into one Earth resources satellite imagery company, Digital Globe. Digital Globe sells 0.3-meter spatial resolution, multispectral (visible and near-infrared) imagery through its multiple distributors and on its website: http://www.digitalglobe.com. Imagery is priced specifically for a customer's situation and varies with location and area of interest.

The constellation of lesser-known Earth resources satellites is large, with multiple nations maintaining systems, including the governments of Argentina, Brazil, Canada, India, Pakistan, Sweden, South Korea, Russia, and the European Union. These satellites have moderate spatial resolution (10–30 meters) and multispectral sensors. Some have radar sensors. Imagery from the Canadian and European programs is available for a fee through Airbus Defence and Space: http://www.geo-airbusds.com.

NASA sponsors several experimental satellite remote sensing programs. Imagery is commonly available without cost over the Internet: https://lpdaac.usgs.gov. Two popular sources of imagery are the MODIS (Moderate Resolution Imaging Spectroradiometer) and ASTER (Advance Spaceborne Thermal Emission and Reflection Radiometer) satellite sensors. MODIS provides multispectral imagery with 250-meter spatial resolution that is useful for large-region applications. ASTER provides multispectral imagery with 15-meter spatial resolution that is similar to Landsat but with increased spectral resolution. Complete, worldwide coverage is not guaranteed from these experimental satellite sensors.

The constellation of satellites is expanding rapidly with the introduction of small, low-cost "cube" satellites. These are often launched into low Earth orbit as groups of satellites that orbit in formation, providing rapid repeat coverage and even video imagery from space. The various systems are designed for multiple scientific objectives, including atmospheric sounding, with only some systems featuring Earth imaging. Some carry visible and near-infrared imaging sensors with moderate spatial resolution of 5–1,000 meters, with one system—Skybox—providing submeter spatial resolution (Skybox 2014).

These are not the large, complex multispectral satellites that characterize national remote sensing programs such as Landsat. Rather, they are simple, solid-state systems built with off-the-shelf components or, in the case of Skybox, custom-designed sensors. These systems are being launched by university research groups and private commercial ventures. It is expected that images will become readily available in the next few years from cube satellites. Limited coverage will likely be available at little or no cost from academic and other not-for-profit organizations, with reliable comprehensive coverage of an area of interest being provided by commercial companies at price points to be determined.

Imagery from aircraft is ubiquitous throughout the world. Aircraft can be flown from low or high altitude to provide moderate-spatial-resolution imagery and very fine spatial resolution imagery, down to 5 centimeters. The most common type of aircraft color imagery available in the United States is from the National Agriculture Imagery Program (NAIP) distributed by the US Geological Survey and the US Department of Agriculture. This program provides 2-meter and 1-meter spatial resolution imagery in four spectral bands (three visible and the near infrared) for the agricultural lands of the United States on an ongoing basis. Imagery is packaged as county-wide mosaics as well as smaller-area mosaics with boundaries corresponding to the well-known US Geological Survey 7.5-minute topographic maps. There is a nominal charge for NAIP imagery when obtained from the federal government on CD. But the federal government and many state GIS programs provide free NAIP imagery over the Internet. For example, the University of California, Berkeley, offers free imagery through its Geospatial Innovation Facility: http://gif.berkeley.edu/resources/naip.html. The California Department of Fish and Wildlife also provides NAIP imagery as a cloud service over the Internet as Esri layer files and in open source formats: http://www.dfg.ca.gov/biogeodata/gis/map_services.asp.

The US Geological Survey also supports a nationwide aerial photography program that provides panchromatic imagery of the United States at 1-meter spatial resolution. The images are mosaicked into standard image products based on the 7.5-minute topographic map. Images are available over the Internet for download as quarter-map segments, four quarters per 7.5-minute map sheet. These images are called *digital orthophoto quarter quadrangles* (DOQQ, sometimes shortened to DOQ), because the map sheets to which the imagery is registered are commonly referred to as quadrangle maps or simply "quads." The federal government offers DOQQ images as a free download or images on CD for a small fee. Many state governments provide free DOQQs over the Internet as well. *Ortho* signifies that the images have been terrain corrected and georeferenced to standard Earth coordinates using photogrammetric methods, as explained in chapter 6.

Recently, companies that provide mapping services over the Internet are also providing imagery. Natural-color imagery is being streamed into various Internet-based map viewers, but spectral resolution is largely limited to three visible bands, displayed in natural color. The spatial resolution of the imagery varies depending on location, down to 0.15 meter in some urban areas. Historical as well as current imagery is provided. Sources include commercial satellites and aerial photography, when available. The Google Earth viewer allows users to view natural-color imagery and upload GIS layers. You will need Google Earth Pro to download images to your GIS and to perform other functions. Google offers both the Google Earth viewer and Google Earth Pro licenses for free over the Internet. Google is also developing Google Earth Engine for display of visible and infrared imagery, including derived products. Esri also provides imagery as a service, commercially through its ArcGIS Online program and ArcGIS for Desktop software. The Esri imagery services include aerial imagery, commercial satellite imagery, and Landsat imagery, both current and historical, visible and infrared. These image data-streaming services eliminate data storage requirements for the client's computer.

Table 4-1 summarizes the most widely available types of Earth images that are easily compatible with GIS software. Acronyms, defined below the table, are commonly used on various websites providing this imagery and are therefore useful as search parameters.

Table 4-1. Widely available types of remote sensing imagery

Source	Sensors/Programs	Spatial resolution (meters)	Spectral resolution
US National Oceanic and Atmospheric Administration	Geostationary and Polar Orbiting Satellites	1,000–10,000	Partial visible (minus blue), NIR, SWIR, Thermal IR
US Geological Survey	Landsat	15–90	Visible, NIR, SWIR, Thermal IR
European Union, Airbus Defence and Space	Pleiades, SPOT, TerraSAR-X, and others	1.5–20	Visible, NIR, SWIR, Microwave
Digital Globe	World View	0.3–5	Visible, NIR, SWIR
US National Aeronautics and Space Administration	MODIS	250	Visible, NIR, SWIR, Thermal IR, 36 bands
US National Aeronautics and Space Administration	ASTER	15–90	Visible, NIR, SWIR, Thermal IR, 14 bands
US Department of Agriculture	Color Camera, NAIP	1–2	Visible, NIR
US Geological Survey	Panchromatic Camera, DOQQ	1	Visible
Esri, Google, Microsoft	Imagery provided as a service, over the Internet	0.5–30	Visible, NIR

Key terms and abbreviations
ASTER: Advanced Spaceborne Thermal Emission and Reflection radiometer
DOQQ: digital orthophoto quarter quadrangle
LDCM: Landsat Data Continuity Mission
MODIS: Moderate Resolution Imaging Spectroradiometer
NAIP: National Agriculture Imagery Program
NASA: National Aeronautics and Space Administration
NIR: near-infrared wavelengths (0.7–1.3 micrometers)
NOAA: National Oceanic and Atmospheric Administration
SWIR: short-wave infrared wavelengths (1.3–3.0 micrometers)
thermal IR: thermal infrared wavelengths (8–14 micrometers)
visible: visible wavelengths (0.4–0.7 micrometers)

Chapter 5
Displaying digital images with GIS software

The greatest advance in realistic photography and electronic imaging has been the incorporation of color into photography in the mid-twentieth century. Human vision is by nature a multispectral imaging system that uses a unique method of combining three spectral bands of image data into one color image in the brain. Humans can distinguish millions of colors (Waldman 1983). Engineers' ability to mimic human color perception has made photography, television, and remote sensing imagery much richer in information content than using only black-and-white (panchromatic) images.

True-color images

The tristimulus theory of color vision was the key to the development of color imaging, first with film and later in digital screen displays (Waldman 1983). Thomas Young, Isaac Newton, and others experimented with light and prisms in the eighteenth century and discovered that all of the millions of colors that humans perceive are contained within white light. Light of different wavelengths, in a continuum between about 0.4 and 0.7 micrometers, is perceived by humans as various colors. This fact can be observed by watching sunlight pass through a prism or any crystal hanging in a window. The prism separates the white light into the rainbow of colors contained within. Thomas Young formulated a theory of color vision in 1801 that was instrumental for the future development of color photography and digital image display: all of the colors that humans perceive can be created by mixing only three additive primary colors: blue (0.4–0.5 micrometers), green (0.5–0.6 micrometers), and red (0.6–0.7 micrometers), as shown in figure 5-1.

The tristimulus theory of color vision is vital to effective image display. Designers of color imaging devices need to provide for the formation of only three colors, and then combine their various intensities in a single image to form all of the colors that humans perceive. For example, a little more red in the red/blue combination produces a reddish magenta. Three electron guns (red, green, and blue) were used to create early color television displays. Today, more complicated display designs are used to form various combinations of the same three colors using liquid crystal and light-emitting diode (LED) displays. The same theory makes color film possible using three layers of photosensitive emulsion to form variable proportions of red, green, and blue color with subtractive dyes. The theory is so ubiquitous that a specialized term, *RGB* (red, green, and blue), is commonly used to describe color image displays.

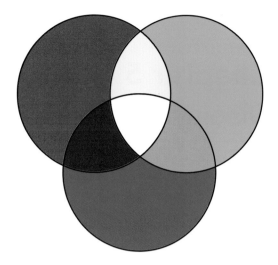

Figure 5-1. A Venn diagram illustrating the tristimulus theory of color vision. The three additive primary colors (red, green, and blue) are shown along with the combinations of equal parts of those same colors. Red and green combine to form yellow. Blue and green combine to form cyan, and red and blue combine to form magenta. All three colors (red, green, and blue) combine to form white.

Figure 5-1 also explains printing technology with subtractive inks. Yellow ink subtracts blue light. It must, by definition, because yellow is made from equal parts of red and green light, no blue. Magenta ink subtracts green light, and cyan ink subtracts red light by the same logic. Imagine a white piece of paper with yellow ink on it. All wavelengths of white light from the sun or a light fixture above the paper are reflected by the white paper up to a human eye. The yellow ink absorbs the blue light, allowing the red and green light to reflect from the paper. The red and green light enters the human eye, and the color yellow is perceived. Humans cannot discriminate equal portions of red and green light from truly yellow wavelengths. If humans look at a spectrally pure yellow, such as a yellow flower petal, the red and green color receptors (cones) of the human eye are equally stimulated, forming yellow in the brain. Again, the tristimulus theory is so pervasive that a specialized term, *YMC* (yellow, magenta, and cyan), is commonly used in printing with subtractive inks. Black ink is then added by printers to embolden the resulting colors.

Assigning spectral bands to colors

To display multispectral images effectively, you must integrate the multispectral concept explained in chapters 2 and 4 with the tristimulus theory, presented in this chapter. The key principle is that you can only display three spectral bands at a time: one in red, one in green, and one in blue. Virtually all known display technology is based on the tristimulus theory. The brightness levels of each spectral band are displayed in various intensities of the three additive colors. For example, if an image obtained in the 0.6- to 0.7-micrometer band is displayed in the red color, areas reflecting a small amount of light in that band would display as very dark red in the resulting color image. Areas reflecting a lot of light in that band would display in bright red. Medium reflectance areas would display in medium red.

This seems like a lot of technical information for the simple display of a color image, and it is true that, for most of us, digital color image display happens automatically. That is, the three spectral bands are packaged into one image file, which makes it easy for us to be unaware that the color image is really made up of three images, one from each third of the visible spectrum. This method for casual photography is intuitive.

However, when there are more than three spectral bands available to view and when some of these bands represent wavelengths of electromagnetic (EM) radiation to which our eyes are not sensitive, we need to understand how spectral bands are assigned to colors and learn to assign them in effective ways to produce false colors. This understanding can also help us modify the color contrast or brightness levels of individual spectral bands instead of applying blanket brightness and contrast adjustments to the entire three-band package. Indeed, understanding tristimulus theory can help you enhance three-band, natural-color imagery and is essential for enhancing false-color imagery.

Natural-color composites

An example of a four-band set of digital imagery with an associated natural-color composite is shown in figure 5-2.

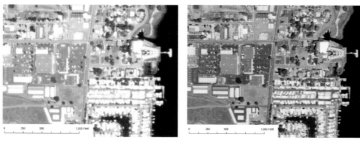

0.4–0.5 micrometer band 1   0.5–0.6 micrometer band 2

0.6–0.7 micrometer band 3   0.8–0.9 micrometer band 4

Natural-Color Composite (band 1 in blue, band 2 in green, band 3 in red)

Figure 5-2. A four-band multispectral image showing the four bands in a grayscale rendition and a natural-color composite image using bands 3, 2, and 1 in red, green, and blue, respectively. West shore of Clear Lake, California, summer 2009.

Courtesy of National Agriculture Imagery Program, US Department of Agriculture, Farm Service Agency, California Department of Fish and Wildlife, Biogeographic Data Branch.

The first three bands of most digital image datasets represent the three additive colors that humans perceive. Band 1 is designated for what humans perceive as blue light (0.4–0.5 micrometers), band 2 for green light (0.5–0.6 micrometers), and band 3 for red light (0.6–0.7 micrometers). The major exception is image data from older moderate-spatial-resolution satellites launched before the turn of the millennium that lack the blue-sensitive band. Band 1 is normally sensitive to green light in these older systems. Also, new systems may depart from this convention of bands 1, 2, and 3 being sensitive to blue, green, and red light, respectively. On Landsat 8, launched in the spring of 2013, both bands 1 and 2 are sensitive to blue light. Band 1 is a short-wavelength blue band (0.435–0.451 micrometers) designed for mapping coastal water and detecting aerosols in the atmosphere. Band 2 (0.452–0.512 micrometers) corresponds to a traditional blue-sensitive band and is used to generate a natural-color composite. Weather satellites and research satellites also depart from a numbering convention that assigns 1, 2, and 3 to blue, green, and red, respectively.

In figure 5-2, the four spectral bands are shown as four black-and-white images, one image for each spectral band. Images from single bands are usually described as *grayscale images* by image analysts because the images are displayed in shades of gray. Normally, the individual spectral bands of a multispectral image are not displayed in gray scale. Rather, three of them (bands 1, 2, and 3) are combined into a single color image using the additive colors of blue, green, and red to form the natural-color composite, thus mimicking human vision. Note how the green trees shown in the color image are slightly brighter (lighter gray) in band 2 (the visible wavelengths that humans perceive as green) than they are in band 3 (visible red). Also note how subtle the tonal differences are between the bands 1, 2, and 3 in the grayscale images. The ability to combine these three bands into a color image greatly increases your ability to interpret the image compared to the three grayscale images. Humans can differentiate many more color variations than shades of gray, so it easier for us to distinguish differences in spectral brightness among Earth surface features presented in color.

Multispectral imaging normally provides more information than the spectral range of human vision. Spectral band 4 in figure 5-2 shows a very different pattern of brightness than the other three bands. Band 4 comes from the near-infrared spectral region, a wavelength band that is invisible to humans yet highly reflected by healthy, living vegetation and absorbed by water. Consequently, water is very dark in this band and vegetation is very bright (see chapter 2). Trees are brighter than their asphalt backgrounds, and the shoreline is enhanced in the near-infrared band. The instrument records brightness levels from this invisible wavelength band that are displayed in shades of gray, which humans do see.

However, when displayed by itself in gray scale, the near-infrared band only highlights the features that have different reflectance percentages in this band. More information about the entire scene is available when the near-infrared band is combined with two of the visible bands to form a false-color composite image.

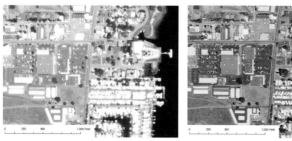

0.4–0.5 micrometer band 1   0.5–0.6 micrometer band 2

0.6–0.7 micrometer band 3   0.8–0.9 micrometer band 4

False-Color Composite (band 2 in blue, band 3 in green, band 4 in red)

Figure 5-3. A four-band multispectral image showing the individual bands in black and white and a false-color composite image using bands 2, 3, and 4 in blue, green, and red, respectively. West shore of Clear Lake, California, summer 2009.

Courtesy of National Agriculture Imagery Program, US Department of Agriculture, Farm Service Agency, California Department of Fish and Wildlife, Biogeographic Data Branch.

Standard false-color composites

Standard false-color composites incorporate the near-infrared band (band 4), displayed in shades of red; the red visible band (band 3), displayed in shades of green; and the green visible band (band 2), displayed in shades of blue (figure 5-3). The near-infrared band contains unique information, so it is well worth the effort to understand and interpret false colors. The blue visible band (band 1) cannot be displayed in this false-color composite because only three colors are allowed by the tristimulus theory.

The standard false-color composite was developed with photographic film during World War II and manufactured for the military by Kodak as camouflage-detection film. The false-color film was designed to differentiate between vegetation and camouflage paint. Real vegetation, having high reflectance in the near infrared, contrasted sharply with the camouflage paint of the era, which reflected poorly in the near infrared. The military use of the film for camouflage detection was short-lived because manufacturers developed infrared-reflective camouflage paint and netting. However, military photo interpreters preferred the film for enhanced definition of water and various kinds of vegetation, which greatly assisted in terrain analysis. The film enjoyed widespread civilian use after the war and throughout the latter half of the twentieth century because of its excellent qualities for general interpretation. This false-color combination was incorporated into digital imagery as the standard false-color composite, or simply a false-color image, and is quite popular, especially for vegetation mapping (Campbell and Wynne 2011). An additional advantage of the false-color image is that it avoids the blue visible portion of the spectrum between 0.4 and 0.5 micrometers, the wavelength range most affected by atmospheric scattering (chapter 3).

Healthy, living vegetation is shown in a reddish-magenta color in figure 5-3 because of its high reflectance in the near-infrared band (band 4), displayed in red. Vegetation has a somewhat high reflectance in the visible green band (band 2), displayed in blue. Vegetation's lack of reflectance in band 3 (visible red, displayed in green) ensures the virtual absence of green for the trees in this composite image in which leaves are healthy and living. The tristimulus theory stipulates that red and blue form magenta, and this combination has more red than blue, thus forming a reddish-magenta color. Because living vegetation is the only feature that is very dark in band 3 and very bright in band 4, it is the only surface displayed in the magenta color in figure 5-3.

Even more interesting is the variation of magenta colors indicated for the different kinds of living vegetation, wetland shrubs on the shore of the lake and broadleaf trees further inland. Image interpreters tend to be more sensitive to changes in magenta color associated with different vegetation types than changes in green color from the visible spectrum. Also, reflectance differences between needleleaf and broadleaf vegetation are more pronounced in the near-infrared band than in the visible bands (Campbell and Wynne 2011). Roads are a bit easier to see in the false-color rendition (figure 5-3) than in the natural color (figure 5-2), and the increased definition of wetland having emergent vegetation in the false-color image is notable. Broadleaf trees stand out clearly in the false-color image but tend to blend in with other dark features, such as rooftops, in the natural-color image.

Admittedly, the improvement of false color over natural color is not overwhelming for general land cover mapping (although it is critical for vegetation-type mapping). Some interpreters feel that the ability to view the Earth in natural colors makes the natural-color composite the better choice. The good news for GIS professionals is that most image acquisition programs (National Agriculture Imagery Program, for example) and image data servers are now providing four bands of multispectral data so that the imagery can be displayed in natural color (band 1 assigned to blue, band 2 assigned to green, and band 3 assigned to red) or in standard false color (band 2 in blue, band 3 in green, and band 4 in red). You can assign any band to one of the three additive colors. With four bands and three additive colors available, 24 color permutations are possible. Only two permutations are normally used, however. The natural-color composite, which mimics human vision, is always popular because the landscape is presented as humans would perceive it if they were viewing from above. Only one false-color composite is normally used: the one in which near-infrared reflectance is displayed in shades of red. This is the combination incorporated into false-color infrared film, which had a successful 50-year history in remote sensing. Experienced image interpreters are familiar with this color infrared rendition, and they appreciate how well it works for vegetation-type identification and water detection.

The color choices increase when more spectral bands are available, as is the case with many Earth resources satellites that incorporate a short-wave infrared (SWIR) band with the near-infrared band and three bands in the visible spectrum.

False-color composites using short-wave infrared

Two false-color composites are popular when using a SWIR band, the near-infrared band and a visible band (figure 5-4).

Three color composites are compared in figure 5-4: *A,* natural color; *B,* false color incorporating the SWIR band displayed in green; and *C,* false color incorporating the SWIR band displayed in red. The natural-color image (*A*) shows more color variation in the lake than the other composites. This is because it does not include the near-infrared band, which renders all depths and sediment conditions of water as very dark. However, the natural-color image (*A*) has generally less color variation among the vegetation types than the other composites. It also has a bluish tint because of the haze effect caused by incorporating the blue visible band in the natural-color composite (the wavelength band most affected by haze; see chapter 3).

A variation of the standard false-color composite that incorporates a SWIR band displayed in green is shown in *B.* This combination has become very popular with forest management organizations because it shows the living vegetation in shades of magenta but with increased color variation in vegetation types. Contrast between dead and living vegetation is increased by inclusion of the red visible band displayed in blue. The boundaries of water bodies are more easily delineated in this color composite compared to the natural-color rendition (*A*) because water has such low reflectance in the near-infrared band and fairly low reflectance in the SWIR band. This ensures a dark blue water color, easily distinguished from the shoreline.

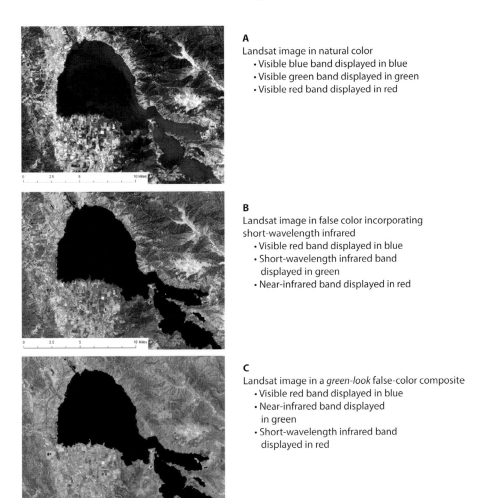

A
Landsat image in natural color
 • Visible blue band displayed in blue
 • Visible green band displayed in green
 • Visible red band displayed in red

B
Landsat image in false color incorporating short-wavelength infrared
 • Visible red band displayed in blue
 • Short-wavelength infrared band
   displayed in green
 • Near-infrared band displayed in red

C
Landsat image in a *green-look* false-color composite
 • Visible red band displayed in blue
 • Near-infrared band displayed
   in green
 • Short-wavelength infrared band
   displayed in red

Figure 5-4. A display of popular color composites used with moderate-spatial-resolution satellite imagery: *A*, a natural-color composite; *B*, a false-color composite incorporating the short-wave infrared (SWIR) band displayed in green; and *C*, a color composite incorporating the SWIR band displayed in red. Clear Lake, California, August 7, 2002.

Landsat images courtesy of US Geological Survey.

Vegetation appears green in the bottom composite (*C*) because the near-infrared band is assigned to green. The high near-infrared reflectance and low visible-red reflectance of living vegetation ensures a green color in the composite. Some organizations prefer this composite because of the more natural-looking green vegetation; but the green color is much brighter than the natural-color image (*A*), and the dead grass is rendered in shades of purple—hardly natural, but readily evident in the image.

The US Landsat program has incorporated a second SWIR band since the mid-1980s to enhance geologic interpretation. Geologists asked for a SWIR band in the clay mineral absorption region of the spectrum, between 2.08 and 2.35 micrometers. Certain clay minerals associated with ore deposits have been identified using this band. This wavelength region is near the long-wave limit for reflected solar radiation (3.0 micrometers). Therefore, it operates with less upwelling radiance than the other bands and is subject to absorption by atmospheric moisture. The image is somewhat darker than images from the 1.55- to 1.75-micrometer band and is therefore not as popular in composites created for mapping land cover or land use. In addition to detecting mineralization, this band has also proven useful in assessing the severity of wildfire effects. The band is used successfully in calculating spectral burn indexes by the US Forest Service.

Using color composite imagery

Table 5-1 summarizes common color composites and provides information on the benefits of each.

Table 5-1. Attributes of common color composite images

Type of color composite	Red color assignment (micrometers)	Green color assignment (micrometers)	Blue color assignment (micrometers)	Benefits
Natural color	Visible red (0.6–0.7)	Visible green (0.5–0.6)	Visible blue (0.4–0.5)	Large variation in color of water, scene looks natural
Standard false color	NIR (0.7–1.3)	Visible red (0.6–0.7)	Visible green (0.5–0.6)	Enhanced delineation of water and vegetation types
SWIR false color	NIR (0.7–1.3)	SWIR (~1.5–1.7)	Visible red (0.6–0.7)	Differentiation of vegetation types
SWIR green-look false color	SWIR (~1.5–1.7)	NIR (0.7–1.3)	Visible red (0.6–0.7)	Living vegetation rendered in green color

Note: NIR = near infrared, SWIR = short-wavelength infrared.

A good rule to follow when creating color composite imagery is to include multiple composites in the stack of GIS layers that are planned for the GIS analysis or mapping project. It is always a good idea to include a natural-color composite (as long as a visible band is available from the blue wavelength range) to provide a rendition of the landscape that mimics what the human eye would perceive. An advantage of natural-color is that different depths and sediment levels in water are more clearly shown in this composite compared to any that include the near-infrared band. Always include a standard false-color composite because of enhanced vegetation discrimination and stark contrast between living vegetation and dead vegetation/nonvegetation. The standard false-color composite also provides good boundary delineation for water bodies because water is so dark. Finally, include a color composite that incorporates a SWIR band (if available), because it often shows more contrast among various land features than either the green visible band or the red visible band. Either a SWIR green look (figure 5-4C) or SWIR false color (figure 5-4B) can be used. You can activate layers appropriate for your audience or mapping task at will in the GIS display when they all are included in the map composition.

How software controls contrast and brightness of color displays

Once the spectral bands are assigned to the additive colors (red, green, and blue), GIS software displays the color composite image. Most software packages automatically enhance the contrast and brightness, and the resulting image quality is adequate in many situations. However, great improvement can be made in some situations by modifying the brightness levels from those assigned by the automatic enhancement. Many GIS software packages now provide a simple contrast and brightness adjustment that is similar to adjustments on a color television. However, most image analysts would argue that you must understand how the software assigns image brightness levels to color brightness levels on the screen for each of the displayed bands in order to most effectively modify image contrast and brightness. Many GIS software packages provide this functionality.

In any digital image, the brightness values of the pixels must be assigned to particular intensity levels of color in RGB display. You might think that task trivial because the image normally has 256 brightness levels, and the color display has 256 levels each of red, green, and blue color. The problem is that seldom is the entire range of possible brightness levels recorded by the sensor. Many images obtained by digital cameras have average brightness levels of about 150, varying from a low value of 50 to a high of 200. The higher brightness levels are only present when imaging snow or clouds or other very bright features. Brightness value ranges are often even narrower for many satellite-based sensors, commonly varying from approximately 0 to 70.

A problem occurs because the computer display needs to be hard wired (that is, unchangeably connected) to the imaging board in order to speed image display. Therefore, for example, brightness level 15 in spectral band 3 must be displayed in color level 15 in the red color (or whatever color is assigned to band 3). The common range of image brightness levels (0–70) should not be displayed as color levels 0–70 as assigned automatically because this fails to exploit the full range of color intensities available in the display. Rather, the range of image brightness values should be stretched out to match the range of brightness values (0–255) of a true-color display. In order for this to happen, the computer needs to know the actual range of brightness values of the original image. This information allows the computer to stretch those values out to make use of the full display range of 0–255 and enhance the display of the image in a way best interpretable by humans. The problem also occurs when modern 11-bit images (having 2,048 brightness levels per spectral band) are displayed with 256 color levels per color. The range of brightness values in the original image data may exceed the range of brightness levels of the display device. Automatically compressing those values without regard to their actual range and distribution will seldom produce optimum brightness and contrast in the resulting image. Twelve-bit-per-color (deep color) displays are available but rare. Even with deep color displays, the distribution of the actual brightness values in the image must be considered as with 8-bit-per-band imagery and 8-bit-per-color displays.

So, the question is, how can you measure the brightness qualities of an image comprehensively without being confused by the dark areas in one part of the image versus bright areas in another part? The answer is to tally the brightness levels of all of the pixels in the image and display the results as a histogram (figure 5-5).

The vertical axis of the histogram shows the number of pixels having a particular brightness value, and the horizontal axis shows the brightness value of those pixels. For example, there may be 3,245 pixels having a brightness value of 45. Those pixels would be represented by a vertical bar on the histogram positioned at the value 45 on the horizontal axis and the value of 3,245 on the vertical axis. Pixels with high brightness levels are shown on the right side of the histogram, dark pixels on the left. The number of pixels occurring at any given brightness value can be large. Images with fine spatial resolution can contain millions of pixels.

The histogram shown in figure 5-5 is from a single-band image that is normally shown in gray scale. Many types of images lend themselves to grayscale display, including single-band radar images and single-band thermal-infrared images. However, it should be remembered that most of the images available for GIS are multispectral images made from three or more spectral bands, and they are commonly shown as a color composite in modern digital displays. Therefore, everything is multiplied by three when it comes to histograms and image display—one histogram for each spectral band displayed in each additive primary color: red, green, and blue. The explanation here will be restricted to one spectral band to keep it simple.

Image from Spectral Band 2 (0.5–0.6 micrometers)

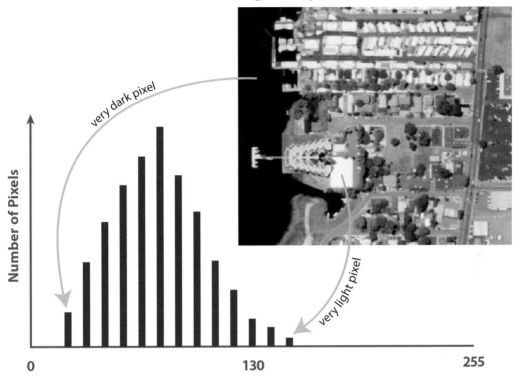

very dark pixel

very light pixel

Number of Pixels

0   130   255

Brightness Value in Spectral Band 2 (0.5–0.6 micrometers)

Figure 5-5. A typical histogram from an image acquired in the green portion of the visible spectrum with a portion of that image for comparison. The image in this figure was extracted from the NAIP image shown in figure 5-2.

Image courtesy of National Agriculture Imagery Program, US Department of Agriculture, Farm Service Agency, California Department of Fish and Wildlife, Biogeographic Data Branch.

Stretching the histogram of a single-band image to enhance contrast and brightness

A single-band image is shown along with its unmodified histogram in figure 5-6A.

The horizontal axes of the histograms in figure 5-6 represent the full range of brightness levels (0–255) available in this spectral band. Each division in the background grid represents 32 brightness levels. The vertical axis represents the pixel counts for each brightness value. The actual count values are not important in this context, only the relative counts, and therefore the counts are not enumerated. The histogram in figure 5-6A shows the distribution of pixels to be

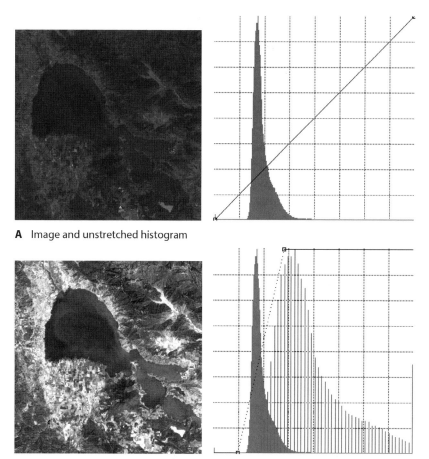

A Image and unstretched histogram

B Image and stretched histogram

Figure 5-6. A portion of a Landsat image from the green visible spectrum with its associated histogram. *A*, The image is displayed with its original histogram. *B*, The histogram has been stretched to increase contrast and brightness.

below the brightness level of 128 (the center of the range) because of the settings of the imaging device. The distribution is also slightly skewed to the left because of the large amount of dark pixels representing the lake. The imaging board displays the original image without stretching the histogram in figure 5-6*A*. All of the pixels in figure 5-6*A* have brightness values below 128, so only darker shades of gray appear on the image display. Remember that the imaging board is programmed to display the 0 level as black, the 128 (middle) level as mid-gray, and the 255 level as white. The image is dark and lacks contrast in figure 5-6*A* because the histogram is not modified, and therefore only the darker shades of gray are used to display the image. You (or an automatic algorithm) must stretch the histogram so that more of the display levels are used by the display device. The enhanced image created from the stretched histogram is shown in figure 5-6B, along with the stretched histogram.

The diagonal lines in the histograms in figure 5-6 represent the linear function used by the software to stretch the histogram. Imagine another vertical axis superimposed on the left vertical axis of these graphs showing the results of the function y = mx + b, where x is the original histogram value, y is the modified (stretched) histogram value, m is the slope of the function, and b is the y-intercept. Both the vertical and horizontal axes are divided into eight sections (each having 32 brightness levels) to assist in the graphical solution to the equation. Find any value on the horizontal axis (x) and move vertically up until the line (linear function) is touched. Then move horizontally across to the left vertical axis to read the modified brightness value (y). Now realize that the new, stretched histogram is laid back down on the original x-axis for comparison with the old, unstretched histogram. Admittedly, that's a lot of imagining, but it does provide an efficient way to graphically portray the results of the histogram stretching function. And, this type of graph is used throughout the image processing and GIS software industry, so it is worth understanding it.

The function runs corner to corner from the lower left to the upper right in the histogram shown figure 5-6A. The slope of the function is therefore equal to 1, and the y-intercept is 0. The linear function reduces to y = x (from: y = mx + 0, y = 1x, y = x). No stretching occurs when the slope is equal to 1 and the y-intercept is 0. Therefore, the new histogram lies directly on top of the old histogram, and it looks like there is only one histogram in figure 5-6A. The purpose of a stretching function is better seen for the condition in figure 5-6B where stretching actually occurs. In this case, the slope is much greater than 1, and the y-intercept is a negative value. Actually, the linear function breaks at y = 0, and the function changes to y = 0 for all x-values below that point because negative brightness values never occur. In the histogram and stretching function shown in figure 5-6B, the brightest original pixels, with a brightness level of about 130, are increased to 255 by the function. To see how this works, find the position of the brightest pixels of the original histogram. Each grid square represents 32 brightness values in the graphs, so the brightest original pixels have a brightness value of about 128 (approximately four squares over from the left zero point). Notice that if one moves up from 128, the line of the function is not reached until the top of the graph, a value of 255. This exercise solves the equation graphically. That is, the original value of 128 is "stretched" to a value of 255. In fact, one can see by the position of the linear function that several brightness values slightly less than 128 are also stretched to 255 because the function breaks flat at y = 255, making all x-values above the break point equal to 255. This results in the large number of pixels on the right of the stretched histogram (shown by the separated tall vertical line on the right side in figure 5-6B). To ensure understanding, note how the mode of the unstretched histogram is equal to about 50 (about one and three-quarter squares over from the left zero point). That point intersects the function at slightly less than 100 on the vertical axis (about three squares up). Therefore, the original brightness value of 50 is stretched to about 100, and you can see the peak of the stretched histogram (the spread out vertical lines) at about 100 (slightly more than three squares over from the left).

Many GIS software packages allow the analyst to define the histogram modification function using graphs similar to those portrayed in figure 5-6. Some also provide a *pick list* of common functions that have proven useful for modifying histograms for effective image display. The method used to define the position of the linear function used in figure 5-6 was to first calculate the mean and standard deviation of the original histogram, then place the lower left corner of the function at −2.5 standard deviations from the mean, and then place the upper right corner of the function at +2.5 standard deviations from the mean brightness level. The function is horizontal above and below these points so that all pixels having brightness values less than −2.5 standard deviations are shown as black (0) and all brightness values above +2.5 are shown as white (255) in a gray-level display. This function is called the *standard deviation stretch*. Another popular method used to position the linear function is to place the lower left corner and the upper right corner such that 99 percent of the brightness values are modified by the function and the darkest half percent are made black and the lightest half percent made white. This is referred to as a *percent clip* because a certain percentage of the brightness levels are clipped from the effect of the function and made white or black.

The graph of these linear functions looks like a ramp (figure 5-6). As a result, the graph is called a *ramp*, and the function producing the stretched histogram is often called a *ramping function*. As with many other specialized terms, the noun has become a verb, as in, "ramp that histogram so I can see some contrast." Care should be taken when modifying ramping functions because a ramp drawn too steeply or in the wrong range can seriously degrade the color quality of an image. Best results are normally obtained using the predetermined functions available from a pick list.

All of this logic, illustrated here with grayscale images, transfers to color image display using three ramps, one for each band displayed in each of the additive colors. The darkest levels are stretched to black, and the brightest levels are stretched to bright red, bright green, and bright blue for each of the three spectral bands displayed in the color composite. Pixels having the very brightest level in all three spectral bands displayed would be given the brightest value of all three additive colors, and the equal combination of brightest red, brightest green, and brightest blue creates white.

One slight complication that is important to understand is that a color display is also used to present grayscale images using equal brightness levels of red, green, and blue to create many shades of gray. It may be less confusing conceptually to switch to a gray-level display when displaying a single-band image. However, that would incur unnecessary cost when the color display can be easily made into a black-and-white display by forcing all colors to have equal intensities. The display is color, and therefore the shades of gray must be created by keeping the three colors equal at each brightness level. Fortunately, the analyst simply selects "gray scale" (or "stretched" in the case of Esri's ArcGIS software) from a menu and specifies a spectral band to display in gray scale while the software ensures equal parts red, green, and blue for each shade of gray.

Virtually all GIS software that displays images provides some kind of default histogram stretch because unmodified images display with such poor brightness and contrast. It is a simple enough task to program a computer to calculate the histogram and perform one of the commonly used stretching functions. These automated methods normally provide reasonable brightness and contrast settings. However, the presence of a large area of white clouds in an image or black marginal areas with large numbers of pixels having a zero brightness value can adversely affect the automatic stretching procedures. It is useful in these cases to inspect the images and the histograms for each spectral band displayed and modify the histograms as necessary to achieve good contrast and brightness. Some GIS software allows the practitioner to change the area of the image from which the histogram is calculated. Contrast and brightness are often improved tremendously by collecting the histogram from a noncloudy portion of the image so that the automatic enhancements can work effectively.

Graphs created by GIS software are displayed only for humans to better understand the process. Images are actually enhanced using the mathematical ramping function and a simple look-up table in order to increase the speed of display. The computer calculates every modified histogram value based on every possible input value using the ramping function. This is a small table because there are only 256 possible values in 8-bit images and 2,048 levels in 11-bit images. All possible enhanced brightness values are quickly calculated by computer, and then millions of enhanced brightness values for the entire image are rapidly looked up in the table. More important to you is that the look-up table can be saved as a separate file and used repeatedly once an optimum stretch is achieved. The look-up table is often called a *LUT*, and the acronym is so commonly used that it has become part of image enhancement terminology.

Pseudo color images

Pseudo color images are based on a single spectral band. Predetermined colors are assigned to different brightness levels in a single-band image to create a pseudo color rendition, sometimes referred to as a *thematic color scheme*. Examples include single-band thermal-infrared images that are color-coded in intuitive ways (figure 5-7).

Figure 5-7. A single-band thermal-infrared (3- to 5-micrometer) oblique image of a steam plant. The image has been color-coded in a psychologically meaningful way so that hotter surfaces appear red, blending to yellow and white with increased temperature. Varying amounts of heat loss can be seen from the steam distribution system buried 5 feet under the street.

Courtesy of AITscan, a division of Stockton Infrared Thermographic Services, Inc. Used by permission.

The colors are formed by applying a color table of predetermined colors to the histogram of the single-band image. Brightness levels 0–10 might be colored blue while brighter levels might be colored red, for example. Modern GIS software provides multiple color tables that can be applied to single-band images. Some GIS software allows the practitioner to create custom color tables. Several variations of color tables are commonly used for different types of image data. Tables that blend one color into another or that blend multiple shades of a single color are sometimes called *color ramps.*

Some refer to these pseudo color images as false-color images because they are not based on the tristimulus theory of color vision, which requires the display of three spectral bands in three additive primary colors: red, green, and blue (RGB). However, the term *false color* used to describe these single-band color images is not recommended because most image analysts refer to a false-color combination of three spectral bands as a false-color image even though the image is truly a color image made from three additive colors. GIS professionals must understand that true-color images can be either natural-color images or false-color images because they are both made from three spectral bands based on the tristimulus theory. False-color images are made the same way that natural-color images are made; that is, three spectral bands displayed in three additive primary colors, just not the ones humans would see with their eyes. Pseudo color images are made from a single band by assigning a color table to that band.

Another type of raster dataset that is not an image is digital elevation data (chapter 1). Digital elevation data are often represented by a raster grid that looks like a single-band image. However, it is unique in that the pixels represent elevation rather than brightness. Color tables have been especially designed for raster elevation data such that the darkest pixels (lowest elevation) are colored blue and the brightest pixels (highest elevation) are colored white. The blue looks like water, and sea level is usually the lowest feature in a digital elevation model. White looks like snow, expected to be on mountain peaks. Elevation data normally appear in a more interpretable form when displayed in a pseudo color rendition rather than in gray scale.

Chapter 6
Generating three-dimensional data with photogrammetric measurements and active sensors

The geomatics industry (including GIS, surveying, and mapping) is keenly interested in all things three dimensional (3D). From local building models to continental-scale surface models, representations of the Earth are increasingly being displayed in three dimensions rather than traditional two-dimensional (2D) cartographic displays (e.g., GIM International 2012).

Traditional maps are two dimensional. These maps represent horizontal distance on a plane (a datum plane) and are therefore referred to as *planimetric*. Cartographers project the curved (and uneven) surface of the Earth onto a flat plane such that all distances on the map represent horizontal distances on the ground. This process is called *orthographic projection* because all of the projection lines are orthogonal to the datum plane. Map projections and geodetic datums are well understood by GIS professionals (IIiffe and Lott 2008) and not further explained here.

The vertical dimension (elevation) can be represented by contour lines on maps. Maps containing contour lines are called *topographic maps*. Cartographers also use topographic shading to represent elevation, often in combination with contour lines. Contour lines are being replaced by three-dimensional perspective views in modern visualizations, and building footprints are becoming building models. These three-dimensional models can include internal details that allow viewers to "move" through buildings and urban environments in realistic virtual representations. Indeed, the three-dimensional virtual representation of the Hometree in the popular movie *Avatar* is not far from reality in the display technology being developed today.

Three major remote sensing technologies support three-dimensional representation of the Earth: aerial photography in stereo, airborne laser scanning (lidar), and interferometric imaging radar. These technologies provide a vertical dimension (z-coordinate) that creates the ability to model in three dimensions when combined with the traditional horizontal x- and y-coordinates of flat-map cartography.

Stereo photography, developed before the invention of the airplane (hot-air balloons were used for aerial platforms prior to the twentieth century), has long supported topographic mapping in three dimensions throughout the world. Its growth and application have been especially strong since World War II. The process of measuring the position of features using aerial photography is called *photogrammetry* (*photo* = "light," *gram* = "picture," *metry* = "measure"; literally, measuring positions on pictures made with light).

Photogrammetry continues to provide fine-spatial-resolution images and good positional accuracy with the supporting navigation technologies of the Global Navigation Satellite System (GNSS) and inertial measurement units (IMU). The GNSS is widely known in the United States as the Global Positioning System (GPS), which is actually only the US constellation of navigational satellites. The Russian Federation sponsors the GLONASS, and the European Union sponsors the Galileo system. Modern navigation receivers commonly receive signals from the GLONASS and GPS operational systems to calculate precise Earth coordinates even though many people in the United States refer to them as "GPS receivers." The acronym *GPS* is used throughout this book to refer to precise satellite navigation from multiple satellite constellations, a subject well known to GIS professionals.

Whereas GPSs mounted in the aerial platform provide the precise x-, y-, and z-coordinate location at the moment the image is acquired, the IMU provides precise information on the attitude of the sensor in terms of tilt, tip, and yaw. Working together, these navigation tools have greatly reduced the need for ground control in order to establish accurate positioning from remote sensing.

Lidar systems generate three-dimensional point cloud data that represent the three-dimensional world. They scan the surface with laser light to record millions of data points, each having an accurate three-dimensional location. Lidar systems blur the distinction between remote sensing and field surveying because they can be operated from aircraft at great distance from the surface of interest or from ground-based platforms only a few feet away from the surface of interest.

Interferometric radar systems that measure the phase shift of returning waves provide precise vertical positioning, useful for measuring surface deformation due to geologic processes. This precision is not associated with high accuracy, however, so elevation data that are generated by IFSAR (interferometric synthetic aperture radar) must be calibrated with ground references to achieve acceptable accuracy.

Obtaining vertical and horizontal positions from aerial photographs

The photogrammetric procedures briefly described here (except measuring relief displacement in a single image) are generally not available to GIS practitioners using GIS software. These methods require specialized photogrammetric software and specifically trained people (McGlone 2013).

GIS professionals use the orthoimages and digital surface models created by photogrammetric methods and should be aware of how these products are generated. You should understand the limitations of these products and be prepared for an occasional double or mismatched image of a well-defined feature or for a surface that departs from true ground level.

Geometry of a single aerial photograph

Aerial photographs provide a two-dimensional point perspective view of the terrain below (Wolf 1983). When obtained with a perfectly vertical view angle, they create a map-like representation of the surface. However, the images are only a map (an orthographic projection of the Earth's surface having a constant scale) if the image is perfectly vertical and the terrain below is absolutely flat (figure 6-1).

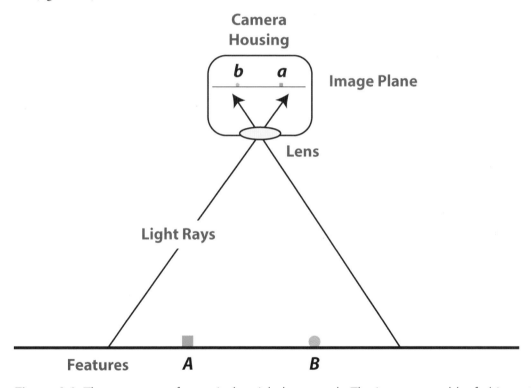

Figure 6-1. The geometry of a vertical aerial photograph. The images *a* and *b* of objects *A* and *B* are portrayed at constant scale across the image plane.

The diagram shows a vertical image of a flat Earth surface containing two features, *A* and *B*. The light rays reflecting from the surface pass through the camera lens and form an inverted image on the image plane. Of course, millions of light rays form the image, but only the outside rays are shown. The ratio of distances between the images *a* and *b* of objects *A* and *B* and the actual distance between the objects *A* and *B* on the ground defines the scale fraction. For example:

10 millimeters image distance/120 meters ground distance

Converting and canceling units:

10 mm/120 m = 10 mm/120,000 mm = 1/12,000, or 1:12,000

The scale fraction is constant across the image if the view is vertical and the surface is flat. It is true that the distance from a feature located at the edge of the image to the camera lens is slightly greater than the distance from a feature located at the center of the image to the camera lens. However, the image of the edge feature is farther away from the lens inside the camera, so diverging light rays of the focusing process render the entire surface at constant scale.

However, the Earth's surface is not flat or smooth, with the exception of a calm water surface, which holds a constant elevation. Even a water surface is curved because the Earth is round. But if a small portion of the Earth is considered, the curvature of the planet affects scale very little. Differences in elevation throughout the image do affect scale drastically, however. The image is a point perspective rather than an orthographic projection because the light rays are focused (cross) through the lens. Differences in elevation produce a displaced location for the image of a point that sits at higher or lower elevation than the center of the image. The truly vertical view occurs only at the center of the image (actually the truly vertical view occurs at the point directly beneath the focal point of the lens, the *nadir* point, but the center of the image is very close to this point when the view angle is very close to vertical).

In aerial photography, the relative altitude of the camera changes as the elevation of the surfaces in the image change. As the relative altitude of the camera changes, so does the scale of the image. This means that photographs of rugged terrain cannot be simply rotated and stretched to overlay precisely on a map. Points positioned at a higher elevation than the image center are displaced outward, away from the center (these points are closer to the camera, producing larger scale). Points positioned at a lower elevation than the image center are displaced inward, toward the center (smaller scale). The displacement of the top of an object relative to its base produces the perceived lean of the images of tall objects in vertical-view imagery (figure 6-2).

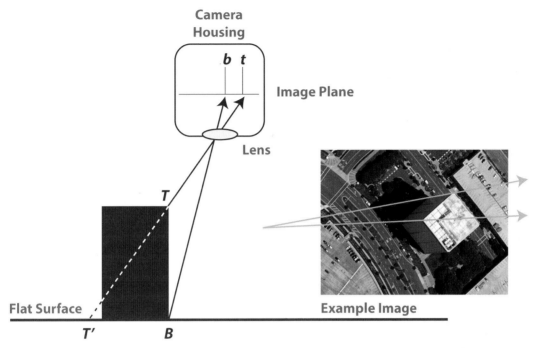

Figure 6-2. A diagram showing relief displacement for a tall building in a vertical-view image. The image of the top of the building (*t*) is displaced radially outward from the image center relative to the image of the base of the building (*b*). The image of the top of the building appears to originate from position *T'* on the relatively flat ground around the building.

Image courtesy of National Agriculture Imagery Program, US Department of Agriculture, Farm Service Agency, California Department of Fish and Wildlife, Biogeographic Data Branch.

If the image were truly map-like (a planimetric surface from an orthographic projection), the building would appear as a building *footprint*, with the top having the same size as the bottom and exactly covering it. The displacement is problematic because detail of the surface near the building is hidden by the apparent lean of the building, and the top of the building is imaged at a larger scale than the base of the building. However, part of the side of the building is shown in the image that would otherwise not be shown in an orthographic projection.

More important, from a regional mapping perspective, is that scale varies as the elevation varies throughout an image because of relief displacement. Generally, the greater the topographic relief, the greater the relief displacement and resulting scale variation for points located at equal distance from the center of a photograph. Also, aerial photographs are never truly vertical, having tilts of up to three degrees in typical aerial survey projects. Therefore, tilt distortion occurs as well as relief displacement.

The height of the building (or any vertical object) affects the displacement. The tops of taller buildings are displaced more than the tops of shorter buildings, other things being equal. The position of the building top relative to the center of the image also affects the displacement; equal-height buildings positioned farther away from the center of the image are displaced more than buildings positioned close to the center. And finally, the altitude of the aircraft above the ground affects displacement, because lower altitudes produce increased displacement, other things being equal. The height of a vertical object can be calculated when the other three variables are known or measured: radial displacement, position of the object top relative to the nadir point (approximated in a vertical image as the image center), and altitude of the aircraft above ground level. The altitude above ground level is normally known when using modern imaging systems because the aircraft is equipped with precise navigation instruments based on GPS and IMU. The center of the image can be obtained from known camera parameters. Altitude above ground level can be calculated from image scale and camera focal length when using older photographs taken without a precise camera location.

The radial displacement and position of the object top relative to image center can be measured and object height can be calculated using a simple formula:

$$h = dH/r$$

Where:
h = object height,
d = radial displacement of the object top relative to its base,
H = height of the camera above ground level, and
r = radial distance from the image of the object top to the center of image.

Units of d and r are the same (usually millimeters), and they cancel out. The height of the object is calculated in the units of H (commonly in feet or meters).

While traditionally the domain of specialized photogrammetric software, at least one GIS software package (Esri's ArcGIS software) has the functionality necessary to make these measurements and calculate object height.

Geometry of an overlapping pair of aerial photographs

Although relief displacement can be used to measure object height from a single, near-vertical image, as previously described, photogrammetrists have also developed a more effective way to measure object heights and surface elevation using stereo pairs of near-vertical images (Wolf 1983; McGlone 2013). The problem with measuring relief displacement in single images is that it can only be done for vertical objects for which the base is visible in the image because the image distance from top to base (the relief displacement) must be measured. Mountain tops and valley bottoms are displaced, as are all points at different elevations than the center of the near-vertical aerial image. However, the base of these features cannot be seen. The base of a sloping hillside (the point directly beneath the top of the hill) is underground, making its location impossible to determine from a single image. Imagine that point T in figure 6-2 is a mountain top rather than a building top. This condition is shown in figure 6-3.

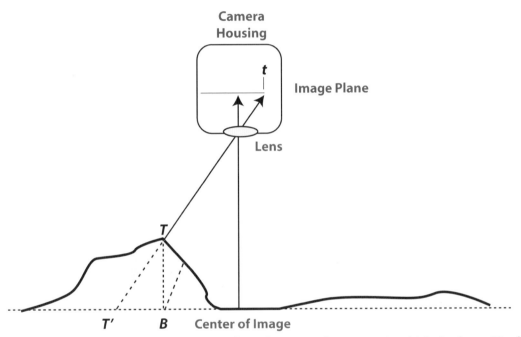

Figure 6-3. A diagram illustrating relief displacement for terrain in which the base (B) of the "object" is underground and invisible in the image.

Point B in figure 6-3 is underground and cannot be seen in the image. The true position of the image of point B is unknown. The image of point T is labeled t, and it is visible in the image. However, t is displaced from its true position so that it appears to be located at T', which is underground and invisible. The height of the mountain cannot be determined from measuring relief displacement in a single image.

A stereo view of overlapping near-vertical images provides the solution to this problem. Two images that overlap in their coverage can be used to generate a three-dimensional stereo "model" of the Earth's surface and derive x-, y-, and z-coordinates for any position within the stereo model. Photogrammetrists accomplish this by measuring the parallax displacement of any location in the stereo model. Parallax displacement is the apparent movement of an object due to the movement of the observer. Objects that are closer to the observer appear to move a greater distance when the observer moves than do objects located farther away. While many people are unfamiliar with parallax in this specific setting of overlapping aerial images, most have experienced the effect of parallax while looking out the window of a car. The fence posts or power poles next to the road appear to go by much faster than the trees or mountains in the distance. The power poles are closer to the car, and therefore their apparent motion is much greater than objects located farther away. Of course the objects are not moving; they only appear to move due to the relative motion of the car. The objects in the near view have greater parallax displacement relative to the observer. That is, they have farther to move relative to the speed of the car and must appear to move faster than objects in the far distance. The moon, for example, appears to be stationary to the observer in the moving car because it is so much farther away from the car than any terrestrial object.

Photogrammetrists measure parallax in stereo pairs of near-vertical aerial images to determine the elevation of any point in the overlapped area (figure 6-4).

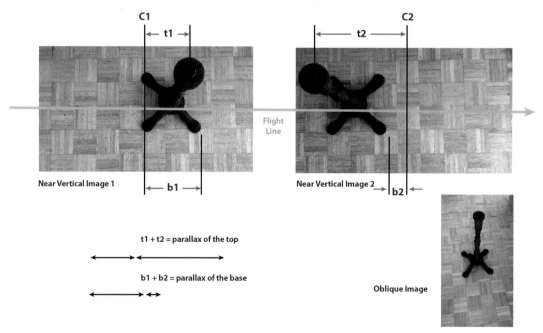

Figure 6-4. A stereo pair of near-vertical images taken with a handheld camera positioned about 4.5 feet above the floor. A candlestick having a height of 2 feet is imaged from two perspectives, approximately 1.3 feet apart, with parallax measurements indicated for the top and base of the candlestick. An oblique image is included in the lower right for orientation.

Parallax displacement is much greater and therefore easier to visualize in this handheld, indoor stereo pair compared to actual aerial images in which the camera is much higher relative to any height variation imaged below. The same principles apply in actual aerial images, but differences in parallax displacement are much smaller. All locations in the stereo overlap zone of two images exhibit parallax displacement. The two camera positions (*C1* and *C2*) represent two positions of the observer, and the distance between exposures is the distance that the observer moved between exposures. The image of the base of the candlestick appears to move relative to the center of each image, from a distance *b1* to the right of the center line in image 1 to a distance *b2* to the left of the center line in image 2. The apparent movement is equal to the sum of *b1* and *b2*, which is the parallax of the base of the candlestick. Similar logic determines the parallax of the top of the candlestick to be the sum of *t1* and *t2*. Parallax equations represent parallax as a signed difference to account for all possible locations within the stereo overlap zone: $+b1 - (-b2)$, which is $b1 + b2$ in this case. The parallax of the top of the candlestick is larger than the parallax of the base, as shown in figure 6-4. Higher elevation points have more parallax displacement than lower elevation points because they are closer to the camera (observer) than positions at lower elevation.

The truly significant thing about parallax is that it can be accurately measured regardless of whether the feature of interest is a building or a mountain. The parallax of any point can be measured in the overlapping pair of images without regard for the position of the image of the base of that object. That is, the elevation of a mountain can be determined without seeing inside the mountain. All points of equal elevation in the overlapping image area have the same parallax. Points located at higher elevation have greater parallax, and points located lower have less parallax. Parallax values form something similar to a series of horizontal slices that pass through the stereo model of the landscape, defining specific elevations.

Even more interesting for photogrammetry is that the overlapped images produce a three-dimensional model of the terrain in the mind when viewed in stereo. While most people are unfamiliar with advanced photogrammetric plotting instruments, many have used the View-Master toy viewer as children or have seen the "hidden image" posters that were so popular in shopping malls many years ago (or at least tried to see them). Stereo perception while viewing an overlapped pair of images is an acquired skill requiring one to relax focus to let the images "go double," thereby allowing the images to drift together into one three-dimensional image. Stereo perception is much easier when looking into a stereoscope or stereo plotter. Operators of stereo plotters compare the relative elevation of various features in the stereo model with the elevation of a floating mark introduced into the stereo model by the plotting instrument. These perceptions in stereo are made much more accurately and quickly than measuring the parallax directly on each photograph with a micrometer. The stereo pair not only facilitates the measurement of elevation in stereo but also enables the plotting of correct horizontal position of any point through the parallax equations (Wolf 1983).

Digital surface models and orthophotos

Modern photogrammetric software produces a digital surface model and an orthorectified image of the terrain from blocks of overlapping, near-vertical images through a largely automated process. The true location of every pixel in the stereo model can be determined once the elevation of every pixel in the image is known from photogrammetic calculations. An orthographically correct image is then created from these corrected pixel locations. These corrected images are often mosaicked together to cover a large area. The result is often termed an *orthophoto mosaic* or, more simply, an *orthophoto*.

Images are not truly vertical in the real world. As a result, the formulas for finding true positions, horizontal and vertical, are complex and not elegant. The process involves systems of complex equations with multiple variables solved through linear algebra. The equations are not linear, and approximations must be made using the linear term of the Taylor series, necessitating calculation of partial derivatives, with solutions converging on the result having the lowest error. Operator intervention is needed occasionally when the automated procedures are fooled by lack of contrast or excessive shadow in some parts of the stereo overlap area of images. Also, dense vegetation can form a "surface" that is above the actual surface of the Earth. Unless an operator intervenes to correct the ground level, the "surface" is formed at an elevation that is above the actual surface. At times the operator has to make a best estimate of the actual surface when vegetation canopies are very dense and the actual ground is not visible through occasional openings in the canopy.

Historically, only relative elevations and relative horizontal positions could be plotted from stereo images unless some true ground coordinates were known. Ground survey was required to determine actual locations of some points, called *ground control points*. Ground control points are obtained from surveying instruments such as total stations and precision GPS receivers. Various methods for extending expensive ground measurements over several stereo models through the process of *aero triangulation* were developed in the late twentieth century (Wolf 1983). Photogrammetrists digitally triangulate the location of points (called *pass points*) that appear in overlapping images during aero triangulation. Adjacent images are fitted to images with actual ground control points allowing the extension of expensive ground control over a block of images. Modern GPS navigation and inertial navigation instruments have greatly reduced the need for ground control today. The position of the camera in the horizontal (x and y) and vertical dimension (z) is known because of precise navigation. The attitude of the camera in terms of tilt, roll, and yaw is also known from the IMU in the aircraft.

Incorporating machine vision into photogrammetry

A recent development in the photogrammetric process is an expansion to hundreds of overlapping images rather than the traditional blocks of 2 to about 20 aerial photographs (Maalouli 2007). Digital cameras require no film, and the cost per image is greatly reduced relative to film photography. As a result, modern aerial surveys are acquiring hundreds of images to cover the same area that would be covered by dozens of film images. Because photogrammetric procedures have been developed to find common tie points among images automatically, little additional time is required to automatically establish thousands of tie points among hundreds of images rather than manually establish tens of tie points over a few images.

Automated photogrammetric procedures are increasingly incorporated into machine vision software. The mathematics is extended to allow the unknown parameters to be excessively overdetermined because many images are available for any tie point or control point. That is, from a mathematical view, the number of equations available is much more than the minimum number needed to determine the unknown parameters. Photogrammetrists have found that the extra measurements improve the accuracy of the procedure so much that they no longer require expensive calibrated mapping cameras. They can achieve necessary precision with inexpensive consumer-grade cameras. The only ground truth required for accuracies of a few meters is an approximate location for each image center. This location is provided by a GPS receiver in the aircraft or unmanned aerial system (UAS). Submeter accuracy is achievable when camera orientation is known from IMU in the aircraft, and a few actual ground control points are obtained with survey-grade ground-based GPS receivers. Also, some aerial survey companies are incorporating oblique images in addition to the near-vertical images, thereby allowing the sides of buildings, trees, or any vertical object to be imaged and eventually incorporated into a three-dimensional model of the scene (Jacobsen 2009).

Obtaining vertical and horizontal positions from lidar sensors

Lidar (light detection and ranging) is an active remote sensing system that employs laser beams to measure the distance from an aircraft to the ground (Vosselman and Maas 2010; Renslow 2012). A laser pulse travels at the constant and rapid speed of light, and the time required for the reflected pulse to return to the sensor is measured very precisely. The time to complete the round trip is divided in half and then used to determine the distance from the sensor to the reflecting surface and ultimately the position of the reflecting surface (figure 6-5).

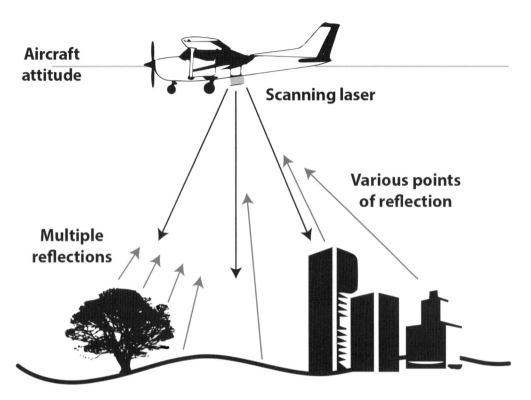

Figure 6-5. A diagram of a lidar system acquiring data from a simple scene. The precise position of the scanning laser is known from a GPS receiver in the aircraft, and the attitude of the aircraft at the instant of the laser pulse is known from an inertial measurement unit (IMU) that records pitch, roll, and yaw.

The laser instrument scans at various look angles and records hundreds of thousands of laser returns per second. The x-, y-, and z-coordinates of the reflection point of each return are calculated based on the position and attitude of the sensor and the time required for the reflected pulse to return. The concept is feasible because of the extremely high speed of laser pulses relative to the speed of the aircraft. Any changes in aircraft attitude or position during the nanoseconds that it takes the laser pulse to return to the sensor are negligible. Therefore, the position and direction of the outgoing laser pulse are known. The coordinates of the point from which the lidar return originated can be determined from the time-distance calculation that determines range from a specific position, and at a specific angle, to the reflecting point. Millions of these points are plotted using computer visualization software to form a three-dimensional cloud of data points commonly called a *point cloud*. Recent developments include multispectral lidar sensors that produce spectral data for each point in addition to elevation data, fusing the concept of a multispectral image with a three-dimensional point cloud.

Multiple returns are recorded for a single outgoing pulse when the area illuminated is semi-transparent, such as a body of water or a vegetative canopy (for example, the tree on the left side of figure 6-5). The instrument keeps track of each return, labeling them as first return, second return, and so on. This capability allows the resulting data to be used to determine water depth up to the limit of laser light transmission through the water. Depths of up to 50 meters have been recorded in very clear water. Vegetation canopy structure and height are often characterized by the recording of multiple returns because the laser beams are only centimeters wide and many of them find their way through even very dense vegetative canopies. Lidar is superior to stereo aerial imaging for finding the actual ground surface under dense vegetation canopies because of these very narrow laser beams. Lidar does not penetrate leaves, however. Therefore, dense tropical rain forest and sometimes dense grass canopies resist the definition of a true ground layer.

Point clouds are displayed using computer visualization software (figure 1-4). The computer power and drive space needed to visualize these large datasets have historically precluded lidar data analysis within a GIS software environment. However, GIS software is increasingly able to display point clouds and convert them to surface models. Traditionally, the aerial survey companies that obtain the lidar data use specialized software and powerful computers to create digital elevation models from the point cloud. These are then imported into GIS software. A grid-format raster dataset is created at a specific density, or *posting* value, that is dependent on the point spacing of the lidar data. The surface reflectance points are then mapped into this grid using their coordinates.

When a digital elevation model is desired, analysts are careful to use only the last return from multiple return systems to increase the likelihood that the elevation model only represents the Earth's surface, rather than trees or other vegetation cover. Lidar analysts also employ advanced algorithms to remove buildings from the elevation model. Elevation models created this way are often called *bare Earth elevation models.* Specific terms have evolved to distinguish between bare Earth elevation models, normally called *DEMs,* and elevation models created from the first returns, normally called *digital surface models,* or *DSMs.* Subtracting the DEM from the DSM generates the approximate height of the vegetative canopy in vegetated landscapes. This result is often termed a *digital canopy model* (DCM). However, there is no guarantee the lidar will capture the very top of the trees.

Lidar visualization software is increasingly incorporated into GIS software packages, allowing analysts to visualize the complete data structure of the lidar point cloud as well as generate and display the DEM/DSM in a three-dimensional visualization. Although lidar represents a very sophisticated technology, its practical application using derived elevation models can be achieved by GIS analysts with only a general understanding of how lidar systems work, as presented here. Further reading is recommended before creating surface models from point cloud datasets (Vosselman and Maas 2010; Renslow 2012).

Obtaining vertical and horizontal positions from interferometric radar sensors

The last remote sensing method addressed here for generating three-dimensional data is interferometric synthetic aperture radar (IFSAR or sometimes, INSAR). IFSAR is a technology so complex and expensive that only large international survey firms or government agencies use it. A well-known IFSAR program was operated from the space shuttle in 2000 (the Shuttle Radar Topography Mission). This joint project with the National Geospatial Intelligence Agency provided elevation data on a near-global scale to generate the most complete, high-resolution elevation dataset of the Earth ever assembled up to that time. The elevation data are available in raster format from the NASA Jet Propulsion Laboratory's Shuttle Radar Topography Mission Data Products web page, http://www2.jpl.nasa.gov/srtm/cbanddataproducts.html.

Radar uses very long (several centimeters) wavelengths of electromagnetic (EM) radiation to image the Earth through clouds and rain using a side-looking geometry, as described in chapter 4. The underlying concept of IFSAR is that the elevation of any feature on the Earth's surface can be established by measuring the phase shift of two radar returns from that feature with two antennas separated by a base distance. The pair of returns is sensed by one radar system having two widely spaced receiving antennae or from two radar systems that pass over the same area in separated flight paths.

The basic principle of interferometric radar is that because the radar pulse is a coherent beam of EM radiation, its wavelength is known and its phase, or wave position, is also known. The elevation of the point from which the radar beam returned can be determined by measuring the phase shift in the return signal from a second antenna located at some distance from the originating antenna. The phase is shifted by the x- and y-location of any point and its elevation, but the shift associated with horizontal location can be removed because the horizontal location is known from the time-distance transformation (chapter 4). Once the shift due to horizontal location is known, the shift due to elevation difference can be measured.

The vertical spatial resolution is theoretically only limited by the wavelength of the radar. In other words, 3-centimeter wavelength radar could conceivably measure elevation differences as small as one wavelength (3 centimeters). This level of precision has been essentially achieved, but accuracy has proven to be much less. That is, the difference in elevation between two points can be measured very precisely, and this quality has facilitated surface deformation mapping associated with volcanoes and earthquakes. However, the actual elevation of points is not determined with great accuracy. As a result, IFSAR surveys must be carefully calibrated with actual elevations determined from ground-based surveys or other forms of remote sensing.

Interferometric radar image processing is so complicated that it is unlikely to become part of the standard tool box for GIS software in the near future. Rather the GIS practitioner normally works with the DEMs produced from IFSAR. Some vendors have developed radar processing tools that work within GIS software, such as PCI Geomatics within ArcGIS (GIM International 2011).

Chapter 7
Image processing

Digital image processing can refer to any computer procedure that modifies digital images. The subject is often divided into three types of activities: restoration, rectification, and enhancement. Image processing is expanded here to include converting brightness values to radiance and atmospheric correction of brightness values. The emphasis in this chapter is on the automated execution of routines that process multiple images with little human input and processes that require image processing software. Contrast enhancement and rectification are also described in chapters 5 and 6 because they are often done manually with GIS software. This chapter includes a typical workflow illustrating image processing procedures that are normally done by image providers, image analysts, and GIS software users.

Image restoration

Processes that modify images in order to improve their appearance and interpretability fall into the category of image restoration. The many procedures of image restoration are mainly focused on images that have been damaged in transmission, interfered with by the sensor system, or purposely degraded by other parties. Although images coming directly from modern remote sensing instruments are usually in near-perfect condition, they sometimes need modification for minor flaws. For example, sensors sometimes drop scan lines, which causes a gap in the image. The gap is filled using the average of the brightness values above and below the gap. This method works well because brightness does not usually vary much between adjacent lines of pixels.

Image restoration techniques can also minimize image "noise." Extremely bright or dark pixels caused by interference in the signal generated by the sensor can be removed using noise-reduction algorithms. These algorithms are based on patterns of pixel averaging. They work well because it is fairly easy to distinguish between noise and very bright or very dark landscape features, which are normally larger than a single pixel.

Filling image gaps and reducing noise in an image are usually done by the organization generating the imagery, either an aerial survey company or Earth satellite data provider. However, it is important for you to be aware of these preprocessing steps. If you are using imagery for quantitative work on very small features that requires working with unaltered brightness values (e.g., estimating surface temperature), you may want to use unprocessed imagery. Most image collection vendors will allow you to obtain unprocessed imagery. However, if you are using imagery in GIS to identify, map, and explore the relative difference in brightness of features that are larger than a pixel, then the standard cleaned-up imagery is desired.

Image compression, sometimes used as a final step of image restoration, is part of the general field of computer file compression. The demand for file compression is great throughout the computer industry, and numerous algorithms have been developed to compress the size of image files. GIS software is increasingly able to compress and decompress images into common formats. In fact, the developing trend is that GIS software manages and displays imagery in its compressed format, therefore decreasing the need to import and convert files to a specific format.

When working with compressed imagery, you should know the difference between *lossy* and *lossless* compression. The JPEG format (.jpg; named after the committee that created the standard, the Joint Photographic Experts Group) is a lossy compression of the original uncompressed image file, which is usually a TIFF file (.tif; tagged image file format). Consumer-grade digital photographs are commonly provided in the JPEG format. Professional-grade aerial images are almost always provided in TIFF format. You should avoid using the JPEG file format in projects that require the full radiometric resolution of the original image.

One lossless compression format that is gaining in popularity is "MrSID" (.sid; multiresolution seamless image database). Developed specifically for image and GIS analysis, MrSID files are easily displayed and managed by most GIS software packages and are currently the preferred image compression format for Earth imagery.

Uncompressed image files in TIFF format are still quite popular for Earth imagery. Hard drive space and computer memory size is increasing exponentially, thus reducing the need for image file compression. A GIS-ready TIFF format called *GeoTIFF* is becoming a standard interchange format for orthorectified Earth imagery. This format includes location and projection information parameters in the file header record. No additional map projection information file is required to display an image in its correct location using GIS software when using the GeoTIFF format.

Image rectification

Rectification involves changing the shape, orientation, and size of the image to fit the map projection of other layers in a GIS. The developing trend is that images designed for display and management within a GIS are orthographically corrected and projected before they are delivered to the GIS organization. These images are called *orthophotos* or *orthoimages*. Orthoimages contain Earth coordinates and projection parameters and display in the correct position relative to other layers in a GIS map viewer. Orthoimages are rectified from their native geometry, which includes variable scale and relief displacement (chapter 6).

Some GIS software packages provide image rectification tools. These tools can help you with certain imagery. For example, historical aerial photographs are often scanned into digital form and need to be georeferenced. Multiple overlapping aerial photographs may need to be trimmed and fit into a mosaic to cover a large area of interest. Most GIS software does not provide a suite of photogrammetry tools and cannot be expected to completely remove relief displacement using a photogrammetric approach. However, many GIS software packages do offer generalized best-fit georeferencing routines. Tools for terrain correction, which effectively remove relief displacement, should always be used if they are available in a GIS software package. Relief displacement will degrade the accuracy of the mosaic significantly in all but very flat terrain. A digital elevation model is required for terrain correction.

The most common method of georeferencing an image and giving it Earth coordinates with GIS software is to find the statistically best fit of a set of reference points that occur both on a georeferenced map (GIS layer) and an image of the same geographic area. These corresponding reference points are called *control points*. Road intersections, stream intersections, property boundary intersections, and other culture features that appear on the image and on the map are desirable control points.

All digital images have coordinates (see the discussion on lines and columns in chapter 1), but they are not Earth coordinates. Regression equations that relate image coordinates to Earth coordinates are developed in both the x (longitude) and y (latitude) horizontal directions using the control points (figure 7-1).

In figure 7-1, the universal transverse Mercator (UTM) map coordinates (y) are a linear function of image line coordinates (x). Statistical regression is used to define the function in which the sum of the squared errors is minimized between the control points and the predicted map coordinates. This method is called a *least-squares* fitting technique. The conventional zero point for image coordinates is the upper left corner of the image. The image is oriented generally north-south in this example. So, an increase in line number (image y-coordinate) corresponds with a decrease in the UTM northing coordinate. The process illustrated in figure 7-1 is repeated for the easting map coordinates using the image x-coordinates.

The control points are normally called *ground control points*, but they need not be obtained from ground surveys. In fact, the control points are normally obtained from a map of the area imaged. The accuracy of the method is thus dependent on the accuracy of the map, and the precision of the ground coordinates is dependent on the map scale and the analyst's ability to place a mouse cursor exactly on a point. A second-, third-, or forth-order polynomial regression technique can be used to decrease the error term compared to using a linear function. However,

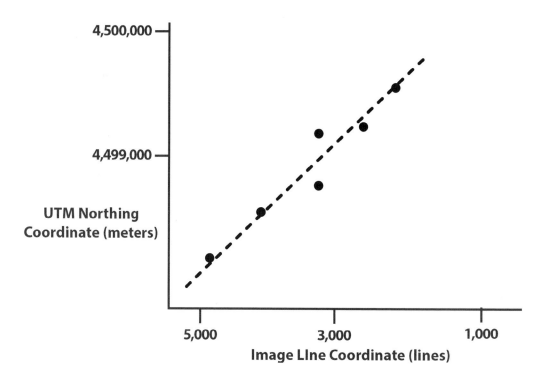

Figure 7-1. A graph of a fitting method to relate image line coordinates to Earth coordinates in the universal transverse Mercator (UTM) projection. The linear regression equation predicts UTM northing coordinates from image line coordinates.

curvilinear functions can depart from the trend in strange ways when the data range is exceeded. Control points should be carefully distributed throughout the range (x and y) of image coordinates to represent the entire image rather than just one area of the image.

Warping functions are used by GIS software to fit the image to the ground control points. A properly oriented and scaled pixel grid is first created from the control points. This grid is then overlaid on the original image pixel grid. The new pixel brightness values are derived from the original values by sampling. Nearest-neighbor sampling fills the new pixel with the brightness value of the pixel closest to the new pixel. Cubic convolution sampling fills the new pixel with a weighted average of the original pixels overlapped by the new pixel. This process reduces the pixel-to-pixel brightness variation in the rectified image compared to the nearest-neighbor approach.

Warping functions can be based on the entire regression equation that applies the same formula to all pixels in the image, or they can use different formulas based on the neighborhood of pixels surrounding a particular control point. The neighborhood method can be more

Essential Earth Imaging for GIS

influenced by a misplaced control point than the comprehensive method, and is therefore not preferred in most cases. Warping techniques based only on least-squares regression do not fully correct for the radial relief displacement of images, and are therefore not true photogrammetric methods (chapter 6). However, some GIS software provides a terrain correction, which is always recommended for removing relief displacement during the rectification process.

Mosaic creation, which has long been the domain of image processing software, is finding its way into GIS software packages. Historically, mosaic creation involved tedious image selection, orthorectification, edge matching, and trimming of overlapped areas by the image analyst. The analyst had to select common points in overlapping images to be used as tie points that ensured a good alignment between images. These tasks are being increasingly automated with the innovation of automatic tie-point identification, automatic color matching, and trimming procedures. These functions do not require analyst input as long as each of the images to be placed into the mosaic is georeferenced and contains the required information in their header records. At least one GIS software suite (Esri's ArcGIS software) creates the mosaic image on the fly, in memory, without creating a second processed image file. Creating the mosaic on the fly avoids the extra space required to create the mosaic and then save the mosaic as a separate large image file.

Image enhancement

Image enhancement includes a variety of functions, from simple contrast enhancement (chapter 5) to advanced image algebra, that produce derived images. Contrast and brightness are often adjusted by the GIS analyst using interactive tools available in GIS software packages (chapter 5). Most GIS software is programmed to automatically enhance brightness and contrast in the image when it is displayed. Some advanced enhancement procedures, traditionally the domain of image processing software packages, are now available in GIS packages, including edge enhancements, multiresolution merging, and one specific spectral band ratio (the normalized difference vegetation index, or NDVI).

Edge enhancements improve the interpretability of linear features and have proven especially helpful for emphasizing roads, urban areas, and geologic fault lines. Edge enhancements should be used lightly. Normally, you would average the edge-enhanced brightness levels with the original image brightness values to create the enhanced version. These subtle nuances are usually programmed into routines, so analysts can generally expect good results with the functions available in most professional GIS software packages.

A multiresolution merge is useful when an image dataset contains several spectral bands at one spatial resolution and one panchromatic band at a finer resolution. The Landsat, SPOT, and Digital Globe satellite images all feature a finer-resolution panchromatic band in addition to the standard-resolution multispectral bands. The process is often called *pan-sharpening* because the finer-resolution panchromatic band is used to sharpen the spatial resolution of the multispectral bands. Various algorithms can be applied to merge the datasets and thereby improve the apparent spatial resolution of the multispectral image. These procedures are traditionally applied by the image provider of commercial imagery or by an image analyst using image processing software. Some of the algorithms used by image providers are proprietary and tend to perform better than methods available in image processing software. Therefore, a multiresolution merge applied by the image provider before the image is input to GIS is recommended. That being said, Landsat imagery is not pan-sharpened by the US Geological Survey before delivery, and some GIS software does provide multiresolution merge functionality. Further reading is recommended for effective use of these techniques (Lillesand, Kiefer, and Chipman 2008; Campbell and Wynne 2011).

The NDVI is so ubiquitous throughout the image processing and GIS world that it deserves special mention here. GIS software can be used to display NDVI images created by image providers or by image analysts, and some GIS packages support calculation of NDVI from multispectral imagery. Land managers have long been interested in detecting changes in vegetation cover over time. The presence or absence of vegetation is often an indication of land use, and vegetative state can be an indicator of environmental condition. The availability of consistently acquired satellite imagery on a predetermined schedule has enabled fairly long-term (decades) repeat coverage with images having identical spatial and spectral resolution. The problem is that varying atmospheric conditions can create changes in image brightness values that are not correlated with changes in vegetation cover. Brightness variations that are not associated with changes on the ground can be reduced by selecting cloud-free images, separated by one or several years, that were acquired during the same season of their respective years. However, atmospheric conditions can still be quite different between images.

The difference in brightness between the digital brightness values of the near-infrared spectral band (NIR; 0.7–1.3 micrometers) and the red spectral band (RED; 0.6–0.7 micrometers) is a good indication of the vegetation condition. For example, dense conifer forests, dense broadleaf forests, and verdant pasture all have high spectral reflectance in the NIR band and low spectral reflectance in the red visible band (chapter 2). Therefore, subtracting the red brightness value from the NIR brightness value (NIR – RED) produces large differences. Dead vegetation, soil, and water have near-equal reflectance in the NIR and the red spectral band. Therefore, they produce much smaller differences (NIR – RED) than vegetated surfaces. While spectral reflectance values are influenced by atmospheric conditions and solar geometry, the ratio of the difference of NIR minus RED, divided by their sum, is much less influenced. Thus, the formula for the NDVI is

(NIR – RED)/(NIR + RED)

Where:

NIR = the brightness value in the near-infrared spectral band, and
RED = the brightness value in the red spectral band.

Smaller differences are divided by smaller sums, and larger differences are divided by larger sums, thus moderating the brightness differences associated with differing atmospheric conditions or different topographic aspects (those differences that are uncorrelated with changes in vegetation cover). The calculation is done for every pixel in the image. A verdant pasture normally has NIR band brightness values of about 40 and red band brightness values of about 5 in an Earth resource satellite image. The NDVI thus calculates as

$(40 − 5)/(40 + 5) = 35/45 = 0.778$

A field of plowed ground typically produces near-equal brightness values in both NIR and red bands. For typical values of about 30, the NDVI becomes

$(30 − 30)/(30 + 30) = 0/60 = 0$

Or, with slight variation

$(30 − 29)/(30 + 29) = 1/59 = 0.017$

Or

$(30 − 31)/(30 + 31) = −1/61 = −0.016$

The NDVI image has only one band (the result of the calculation), with values ranging from +1 to −1. Division by zero is undefined mathematically, yet it can occur for very dark surfaces in which both spectral brightness values are zero. The NDVI value is forced to zero in this case.

The result of the NDVI calculation is shown in figure 7-2.

Note that the NDVI image (*A*) shares the attributes of both of the images that went into the calculation. That is, the highway on the left of the lake is visible, and the agricultural fields are prominent in the NDVI image. The red band image (*B*) provides most of the agricultural field contrast because of verdant and dry pasture/crops in various fields. The near-infrared band (*C*)

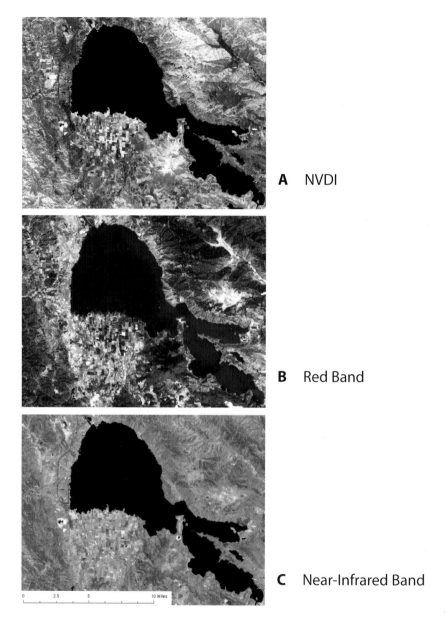

A   NVDI

B   Red Band

C   Near-Infrared Band

Figure 7-2. A comparison of the NDVI (*A*) with the red visible band (*B*) and the near-infrared band (*C*).

Landsat image of Clear Lake, California, on August 7, 2002. Courtesy of US Geological Survey, NDVI created by author.

provides most of the contrast for the darker highway. So, the NDVI reduces the amount of image data from two bands to one band while retaining desirable attributes from both bands.

NDVI images from different dates have less brightness variation due to varying atmospheric conditions and solar geometry than the original images. Because of this, NDVI images are used to detect changes in landscape. A common method of change detection is to subtract the older NDVI image from the more recent NDVI image. Large positive differences in NDVI represent land areas that have become "greener"; that is, verdant vegetation cover has increased over the time period. Large negative differences represent land areas that have become less vegetated, typically associated with land clearing for agricultural, commercial, or residential development.

Conversion to radiance

Actual radiance amounts upwelling from the surface, in watts per steradian (a unit of solid angle) per square meter, can be calculated for each pixel from the brightness values in an image. Conversion factors are published for many sensors by the organizations that acquire the image. These can be used to calculate radiance from image brightness values. Some GIS software supports these calculations. This process of calculating the reflectance percentage or the amount of emitted radiation has been referred to as *quantitative remote sensing* (Liang 2004) and has become increasingly used by remote sensing specialists. The calculation of reflectance percentage is useful for detecting changes of Earth surfaces over time as brightness variation caused by incoming radiation variation is removed. Also, spectral reflectance percentage can be compared to library spectral signatures to determine surface characteristics.

Incoming solar radiation can be modeled accurately as it arrives at the top of the atmosphere given the date and time of day. The ratio of incoming radiance to upwelling radiance defines the percent reflectance as measured at the top of the atmosphere from satellite-based sensors. Conversion to radiance and reflectance percentage is normally accomplished with image processing software before the image is delivered to GIS practitioners. However, some GIS software does support these conversions.

Atmospheric correction

Reflectance percentage or emitted radiance as measured at the top of the atmosphere may be affected by the atmosphere and therefore not represent the actual reflectance or emittance from the Earth's surface. The effects of the atmosphere can be removed through atmospheric modeling of those effects and, once known, removal of effects. The atmosphere attenuates the radiation at different rates at different wavelengths. The atmosphere also scatters radiation to variable degrees, depending on wavelength (chapter 3). Atmospheric scattering adds radiation (path radiance) to the measured brightness of the surface and is most influential in the short-wave portion of the visible spectrum at wavelengths less than 0.5 micrometers. Atmospheric attenuation subtracts radiation from the measured brightness throughout the spectrum. Atmospheric attenuation is especially strong near specific absorption zones caused by water vapor and various elements that make up the atmosphere.

If all you need is the appearance of atmospheric correction, there are a few simple solutions. You can simply adjust brightness levels to match one image to another image. This process facilitates comparison from two dates but does not really correct for atmospheric effects. Also, image providers are now delivering mosaics composed of hundreds of images with brightness levels that have been matched or edge matched. This cosmetically improves the mosaic because colors are more even across join lines.

Minimal or cosmetic atmospheric correction is inadequate when you are interested in the temperature of the surface or when you wish to determine the percentage of reflectance from the surface. Standard atmospheric models are used to estimate scattering and attenuation when radiance and reflectance need to be calculated. GIS professionals are not usually involved in quantitative remote sensing but do use the processed images to more accurately map changes in surface features than is possible with uncorrected imagery. Atmospherically corrected imagery is increasingly available. For example, the NASA Landsat Ecosystem Disturbance Adaptive Processing System (LEDAPS) provides atmospherically corrected satellite imagery for North America (NASA 2013).

A problem with atmospheric modeling is that the atmosphere is variable and seldom behaves exactly like the "standard" atmosphere used in the models. The models have been tested under high solar zenith angles with clear sky conditions. Modeled results compare well with field measurements under these conditions (Liang 2004). However, variable haze conditions and relative humidity may make the model's predictions inaccurate. Atmospheric conditions are seldom measured by weather instruments mounted on weather balloons for the area imaged during image collection. One factor that allows atmospheric modeling to better fit actual conditions is that atmospheric effects are normally constant throughout an image. An image is acquired on a certain date, at a specific time, over a very brief period of time. The atmosphere does not change appreciably over this brief time period. If an analyst can find standard surfaces of known reflectance in the image, the results of the model can be empirically corrected from the difference between the measured reflectance after correction with the standard model and the known reflectance of the surface. In this way, atmospheric models are calibrated for the atmospheric conditions that must have been present at the moment of image acquisition.

Atmospheric correction and radiance calculations normally require image processing software managed by an image analyst. You should be aware of the various stages of the process described here so that you understand what to expect from the image you are interpreting. For example, Landsat 8 images can be processed to convert brightness values to radiance at the top of the atmosphere using information published with the image. This is a radiance value that can be converted to reflectance or surface temperature depending on the wavelengths involved. However, this radiance contains atmospheric effects that can alter the values.

When atmospherically corrected imagery is not available from remote sensing specialists, dark features found in the image can be used to simply subtract the brightness contribution from atmospheric scattering, without the complexity of using an atmospheric model. This process, called *dark object subtraction*, provides a less complex (and less complete) correction of atmospheric effects than using atmospheric models of scattering and transmission. However, it often serves for detecting changes over time as it reduces variations in path radiance, the major contributor to brightness variation in images from multiple dates. Most GIS software packages support this process. An analyst first finds a surface feature (often a deep water body) known to have essentially zero reflectance in a specific spectral band. Because its brightness value should be zero based on reflectance alone, the brightness value recorded in the image must be generated from atmospheric scattering. Therefore, the brightness value of that surface is subtracted from all pixels in the image in that spectral band. This gives the water body a zero brightness value and adjusts the brightness value of all other surfaces to be less affected by atmospheric scattering. Another technique employed in change detection is to select the image having the lowest amount of atmospheric scattering, select a few invariant surfaces in that image, and then modify the brightness values of the other images to match the "clear" image.

Image processing in the cloud

Cloud computing is the use of computer processes that are provided as a service over the Internet rather than being resident on a local server computer or desktop. The cloud consists of many servers that are scalable in terms of computing power and data storage capability. The cloud is especially useful for automated processes that are repetitive in nature and consistently applied. Image processing is increasingly automated, and the amount of imagery available for GIS is growing, requiring massive storage. The cloud-based server environment, with its almost limitless data storage and computer power, provides a flexible platform for processing and delivering very large datasets. Of course, the Internet connection must have sufficient capacity (bandwidth) to download large amounts of image data quickly.

Imagery is increasingly being offered over the Internet as an image service. In these services, GIS software is connected to the cloud, and only the imagery needed for the current view extent is streamed to the client. Examples of these image services are those provided by the California Department of Fish and Wildlife and Esri. Automated photogrammetric processing to produce orthoimages and digital surface models from customer-supplied overlapping images is being offered over the Internet as well (e.g., see the DroneMapper website, http://dronemapper.com). An example of a complete GIS cloud platform is Esri's ArcGIS Online, which offers some GIS functionality (GIS as a service) as well as map publishing over the Internet. All of these services are dependent on high-bandwidth Internet connections. Anything slower than the top tier will provide disappointing performance.

Typical workflow for image processing

As a GIS professional, you will usually be involved in only some of the steps taken to process images (figure 7-3).

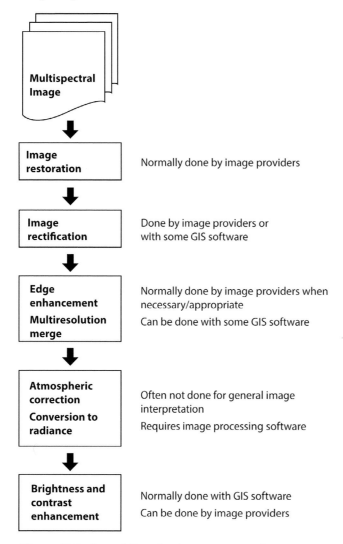

Multispectral Image	
Image restoration	Normally done by image providers
Image rectification	Done by image providers or with some GIS software
Edge enhancement **Multiresolution merge**	Normally done by image providers when necessary/appropriate Can be done with some GIS software
Atmospheric correction **Conversion to radiance**	Often not done for general image interpretation Requires image processing software
Brightness and contrast enhancement	Normally done with GIS software Can be done by image providers

Figure 7-3. A workflow for image processing.

The first step is to restore any defects in the image. This is normally done by the provider of the image data. Next, the image should be geographically rectified. Increasingly, this is also done by the image provider. GIS software can be used to rectify images when they arrive without Earth coordinates. Terrain correction is recommended. An edge enhancement or multiresolution merge is normally done by the image provider when appropriate or necessary. Atmospheric correction using mathematical models normally requires remote sensing specialists using image processing software. Image brightness values can be converted to radiance values using image processing software for quantitative remote sensing. Such conversions are normally not done for general image interpretation in a GIS. However, some GIS software packages do provide enough image processing functionality to convert brightness values to radiance with sensor parameters provided by the organization acquiring the imagery. Finally, brightness and contrast are normally enhanced with GIS software, either automatically or manually. Enhancement can be done by the provider as well, especially if they have created and delivered a mosaic of many digital images.

Chapter 8
Extracting information from images

Modern GIS software provides tools you can use to convert features visible in a georeferenced image into digital information that you can edit and analyze. For example, you can represent small features as points, narrow features as lines, and large features as polygons. This process is called *heads-up digitizing* because the analyst draws points, lines, and polygons on top of the image while looking up at the computer screen. You can use these tools to manually extract new information from the imagery and incorporate it into the GIS.

Some GIS software packages include tools you can use to perform both manual and automated feature identification procedures, including algorithms that automatically classify individual pixels into categories based on similar spectral brightness values. However, the most sophisticated image classification tools are available only in image analysis software. Segmentation procedures that work within regions of the image (object-based image analysis, OBIA) are implemented with image processing software that is used separately from, or sometimes added to, GIS software.

In this chapter, you will learn about manual image interpretation and feature delineation to prepare you to perform these tasks using a GIS software package. You will also learn the advantages and disadvantages of automated image classification methods and how to evaluate maps generated using these methods. Additional reading and practice is required for proficiency in image classification procedures. Lillesand, Kiefer, and Chipman (2008) provide an especially detailed description of image classification procedures, and Blaschke, Lang, and Hay (2008) describe object-based image analysis.

Manual identification and delineation of Earth features using imagery

Image interpreters follow a systematic, rational thought process when identifying features in an image. They methodically check a series of elements that help them consider several attributes of features as they interpret them (the *elements of image interpretation*). This method has increased the probability of correct identification of Earth features over the 100-year history of air photo interpretation. Colwell (1954) pointed out that "using a convergence of evidence brings to bear on a particular interpretation problem all pertinent kinds of information derivable from the image and from outside sources such as maps and tabular information (tide tables for example)."

Elements of image interpretation

Methodically checking the elements of image interpretation is facilitated by GIS, which allows interpreters to work with mapped information and data tables in the same space. This approach to image interpretation is also technology independent (Olson 1960). Innovations in sensor design do not diminish the effectiveness of this approach.

The elements of image interpretation, as defined by Olson (1960), include the following:

1. **Shape.** Shapes can be specifically diagnostic (e.g., the Pentagon building in the upper right corner of figure 2-7) or suggestive of general categories (e.g., the objects with right-angle corners in figure 2-6 that indicate human-built structures).

2. **Size**. The size of a feature can help an interpreter identify what it is. For example, the large size of the oval track in panels *A* and *B* of figure 4-4 suggests an auto racetrack rather than the standard quarter-mile running track. The shape of this feature is less elongated than a normal running track, which also helps with correct identification.

3. **Tone and color**. Tone measures the relative brightness of features in a single-band image. Color represents relative brightness in three spectral bands. The magenta color of the trees in figure 5-3 confirms the interpretation of verdant vegetation in this false-color composite.

4. **Shadow**. Shadows can provide a profile view of vertical objects, which aids interpretation, but they can also obscure surfaces that hinder interpretation. The shadows of the tall buildings in the top and right of figure 1-1 show that they are taller than the buildings in the shopping center. Their height indicates that they are office buildings rather than retail shops.

5. **Pattern**. Regular patterns are characteristic of human-created features. Agricultural row patterns are often diagnostic for various crops, and the spacing of rows can indicate crop type. The row pattern in the lower right of figure 3-2*C* indicates an orchard rather than a natural stand of trees. The ragged nature of the pattern indicates poor maintenance, a sign of abandoned orchards.

6. **Texture.** The appearance of texture or the lack of it in an image is often helpful for correct interpretation. The clumpy texture of the trees on the right side of the color image in figure 5-2 helps to differentiate them from the smooth lakeside shrubs in the same image.

7. **Site**. Certain features are exclusively associated with specific landscape conditions, such as marsh vegetation occurring in low, flat areas. The asphalt parking areas in the color image in figure 5-2 are normally associated with commercial land use, which helps to identify the buildings as commercial.

8. **Association**. Some features are so commonly associated with each other that their presence together can inform a correct interpretation. The bridges and boat docks shown in figure 2-7 are commonly associated with water, and that association helps to interpret all three features. The idea of association speaks to the philosophy of convergence of evidence. In other words, does the evidence converge on what is most likely the correct interpretation?

9. **Resolution**. Spatial resolution determines the level of detail present in the image and therefore influences all of the other elements. For example, a pattern at fine resolution becomes a texture at course resolution. Before reaching a conclusion about the identity of a feature, ask yourself, "Is this texture consistent with what I would expect to see at this level of spatial resolution?"

Checking these elements before arriving at a particular interpretation will help you correctly identify landscape features. Let the evidence converge on the correct interpretation rather than jumping to a quick conclusion based on intuition alone.

Digitizing Earth features using GIS

Once you correctly identify a landscape feature, you can delineate it on the screen using heads-up digitizing. GIS software usually includes tools that allow you to draw points, lines, and polygons using your mouse or stylus. Most GIS software packages also include settings that help you digitize features correctly. For example, your GIS software may include a snapping setting that automatically snaps the end point of a polygon to the beginning point. This setting ensures that your polygon is closed, which saves you a lot of time correcting gaps or overlaps and ensures proper area calculations of your polygon. Most GIS software packages also allow previously drawn boundaries to serve as boundaries of new polygons, which saves drawing time and creates cleaner data.

How scale affects digitization

Consider the expected display scale of your final map product when digitizing features on images. Items digitized at a one scale can have a "broken-glass" appearance when enlarged. For example, if you digitize a near-circular 5-acre lake on an image displayed at medium scale (e.g., 1:24,000), you may create a circle with five to eight vertices. At this scale, your polygon has a diameter of about 0.25 inches, about the size of the hole made by a three-hole paper punch. At this scale, the polygon boundaries appear smooth and rounded. However, if you increase the scale of the polygon to 1:6,000 (four times enlargement, a typical zoom-in with an interactive GIS), the polygon is now about an inch in diameter on screen. At this scale, the polygon boundaries may look flat sided, which may make people think the digitizing is inadequately detailed.

Curved lines are particularly susceptible to this phenomenon. Curved lines do not exist in vector GIS, only the appearance of curved lines when the number of vertices is sufficient to make the small straight-line segments virtually invisible at the display scale. Therefore, to preserve the appearance of curved lines, make sure the display scale of your final map product is smaller than the scale at which it was digitized.

For the best results, digitize features at a scale that allows you to see the full spatial resolution of the image on the screen. Digitizing at this scale will maximize the spatial information content of the resulting map. Therefore, digitize finer-spatial-resolution images at larger scales to capture all of the detail present in the image. Table 8-1 shows what scales to use when digitizing images of varying spatial resolutions. The table includes only the most common sources of imagery for the United States. More detail on these image sources is provided in chapter 4.

Table 8-1. Appropriate scales for digitizing features

Spatial resolution of image on the ground	Typical image source	Appropriate display scale for digitizing features
15–30 meters	Historical national satellite programs: Landsat, SPOT	1:40,000 to 1:60,000
1.5–5 meters	National satellite programs: SPOT and others	1:10,000 to 1:30,000
1–2 meter	DOQQ and NAIP digital aerial photographs, commercial satellite images	1:5,000 to 1:8,000
0.3 meter and finer	UAS images, aerial photographs, and commercial satellite images	1:3,000 and larger

Key terms and abbreviations
DOQQ: digital orthophoto quarter quadrangle
Landsat: US Earth resources satellite program
NAIP: National Agriculture Imagery Program
SPOT: French satellite remote sensing program
UAS: unmanned aerial systems

The area to be digitized and the planned use of the map can help you determine an appropriate scale for digitizing. For example, mapping an archeological dig at 1:500 could be supported nicely with 5-centimeter resolution, low-altitude aerial photographs taken from a small unmanned aerial system (UAS). An appropriate scale for digitizing this fine-resolution imagery would be 1:300.

At some point, you may work on a mapping project where the spatial resolution of the imagery used is finer than the mapping scale requirements. For example, say you are using 1-meter spatial resolution aircraft imagery to produce a regional vegetation map. You acquired the imagery for free over the Internet from a government agency. For this regional application, the map is usually displayed in the GIS at scales of 1:100,000 or smaller. A square, 40-acre agricultural field measures about 0.16 inches on a side at the project mapping scale. If you manually digitize the image at 1:50,000 scale, you could produce an adequate number of vertices and polygon detail for display at 1:100,000. You could digitize the imagery at a much larger scale because the ground resolution is 1 meter, preserving viewing quality at scales as large as 1:5,000. However, digitizing at that larger scale would add to the time required for digitizing and create excess detail in polygon boundaries that could not be appreciated at the planned final display scale of 1:100,000.

How minimum mapping units and map detail affect digitization

When manually digitizing, you should be aware of two things: (1) the *minimum mapping unit* (MMU) of the project and (2) the level of detail required in the final map.

A minimum area for a single polygon is often specified for thematic mapping projects to prevent the final map from being excessively detailed or to comply with government regulations for land mapping in a particular jurisdiction. For example, many vegetation mapping projects specify a minimum polygon size of between 1 and 10 acres. That minimum-sized polygon should be about a half inch across when viewed on screen for manual digitizing to ensure that enough vertices are being digitized to give the final map a realistic appearance.

The concept of proper detail is difficult to define. All boundaries in nature are infinitely detailed. Because maps have been drawn by hand for thousands of years, humans tend to expect a certain level of detail when viewing a map in order for it to be judged as professionally drawn. Too little detail produces a cartoon-like appearance. Too much detail and the digitizing process becomes excessively tedious and time-consuming and the map looks messy and visually unpleasing. The three examples shown in figure 8-1 illustrate this point.

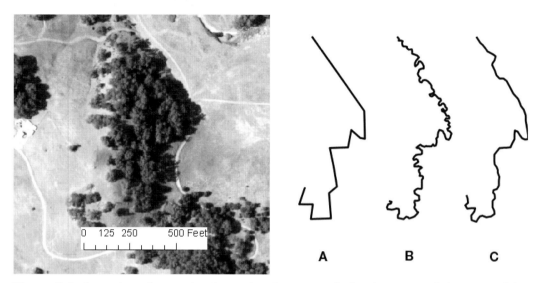

Figure 8-1. A portion of an orthophoto showing a stand of oak trees and three possible boundary interpretations: *A*, *B*, and *C*.

Image courtesy of National Agriculture Imagery Program, US Department of Agriculture, Farm Service Agency, California Department of Fish and Wildlife, Biogeographic Data Branch. Boundaries drawn by the author.

In figure 8-1 the boundary shown in *A* is too generalized. *B* is too detailed. *C* is appropriate for most vegetation mapping work. The scale as shown is larger than one would expect for typical regional mapping projects. The boundary at detail level *C* will downsize nicely. The boundary at *B* will overlap on itself when scale is reduced, causing a messy appearance at the final mapping scale. All of the levels of digitizing detail will look like shards of broken glass if enlarged too much. Coordinate with project managers and colleagues to determine an appropriate level of detail for boundary delineation on a specific mapping project.

One final note on accurate delineation is that you must be aware of shadows when delineating features that produce them. Note that the boundaries drawn in figure 8-1 avoid the shadow area and therefore correctly delineate trees from open areas, not dark areas from light areas. Placing the boundary at the tonal change will include misidentified grassland in the forested polygon. Correct interpretation requires the reliance on texture as well as tone.

Automated delineation of Earth features using imagery

For about 40 years, image analysts have been developing automated methods of image interpretation to speed up the interpretation process and make it more objective. Human interpreters are excellent at integrating multiple factors for identifying objects and land cover classes, but they are slow compared to automated methods. Humans might also overlook small, rare features while scanning through an image. Automated methods that evaluate every pixel will always find pixels of a specified brightness value or that have a range of brightness values even if only a single pixel meets the criteria. The same can be said for automated methods that seek out small groups of pixels. Two or three contiguous pixels could easily be overlooked by a tired human interpreter.

Automated image interpretation has been commonly implemented on a per-pixel basis (called *image classification*) since the 1970s after the launch of the first US Earth resources satellite, Landsat. Several image classification procedures that classify pixels into categories based on their spectral signatures are recognized in the remote sensing literature (Lillesand, Kiefer, and Chipman 2008). Each pixel is evaluated according to its brightness values in several spectral bands and is placed into classes having similar values (figure 8-2).

Classification results are normally color-coded to indicate class membership for each pixel. This is a thematic map, yet it is created with much more detail than one would expect from manually outlining similar surface features into a polygon map. Each pixel of the raster format map has been classified independently, so that map takes on a salt-and-pepper appearance. Some feel that this detracts from the usefulness of the resulting maps. Others argue that a pixel-based raster map better represents the actual diversity of a landscape than polygons drawn by hand.

In any case, per-pixel classifiers work best when the individual pixels are large enough to provide an integrated average reflectance pattern of the contents of the pixel. An individual tree is composed of sunlit crown and shadowed crown, and a stand of trees includes crown shadows as well as sunlit portions. The land cover class "Forest" has little meaning when pixels are smaller than an individual tree crown. One small pixel could be classed as sunlit upper crown, another as shaded lower crown, and another as shaded background. The pixel must be large enough to include several tree crowns that produce an integrated average reflectance of a section of canopy for the Forest class to have meaning spectrally at the pixel level.

Early classification methods

The only spatial resolutions available from satellites before the turn of the millennium were moderate: between 10 and 30 meters. These sensors produced pixels large enough to provide a representative spectral signature for general land cover or land use classes. As a result, many land cover mapping projects were completed using satellite imagery and pixel-based image

Figure 8-2. A comparison of a false-color infrared satellite image on the right side with a classification of that image on the left. Spatial resolution is 30 meters, and the area shown is 7.2 miles across (Clear Lake, California). Greens = various types of tree cover, teal = verdant pasture, yellows = shrubs, gray = barren, blue = water, and white = unclassified.

Landsat image courtesy of US Geological Survey; classification by author based on data from Fox and Garrett (2002).

classification throughout the 1970s, 1980s, and 1990s. Some land managers felt these thematic maps were useful for regional planning at scales smaller than 1:50,000. Others were disappointed by the accuracy of these maps, which averaged between 80 and 85 percent correct. Although an error rate of 15–20 percent seems high to many people, the image analysts were quick to point out that the accuracy of polygon maps drawn by hand, from aerial photographs, was seldom assessed during the 1970s and 1980s. The accuracy of large institutional, manual mapping programs is now assessed in some cases. For example, the accuracy of the US National Wetlands Inventory often exceeds 90 percent.

Early classification methods were limited to simple rules-based algorithms by the weak computers of the day. For example, pixels were placed in a particular class if their brightness value in spectral band X was between value a and value b, with a specific rule for each spectral band.

As computers became more powerful, probabilistic methods were introduced. These methods performed better than rule-based methods because they account for the random variation of pixel brightness throughout an image and for the interaction between spectral bands (covariance). Many per-pixel classifiers are used today. The most commonly used probabilistic method is the *maximum likelihood* approach. This algorithm places an unknown pixel in the class in which it most likely belongs based on statistical tests. Numerous other classification methods have been proposed and implemented, including *neural networks* and *decision trees*. These methods may increase classification accuracy to greater than 80–85 percent when remote sensing imagery is combined with nonimage GIS layers such as soil type, elevation, slope, and aspect (Lu and Weng 2007).

The supervised classification approach

One approach to classification is called *supervised*. This process is called *supervised* because the analyst is closely involved in training the classifier and improving the outcomes of iterations of the classification. When using the supervised approach, the program must know the spectral values (and other attribute values, if they are present in the dataset) of the training pixels, regardless of the algorithm used to classify the unknown pixels. *Training pixels* are a subgroup of pixels (typically less than 1 percent of image pixels) identified as training data to teach the classifier to recognize the attributes of the desired classes. Once trained, the classifier proceeds through the entire image, classifying all of the pixels in the image, including the pixels used for training.

You can modify the procedure to perform the classifications iteratively using new training data to increase classification accuracy using sophisticated image processing software. You do this by changing the probability thresholds for inclusion of an unknown pixel into a class when using statistical methods so that some pixels remain unclassified. Once identified, you can determine the attributes of the unclassified pixels and create a training class or classes for them. You can then reapply the classification algorithm with the new classes included in the training set to increase the accuracy of the classification and the percentage of pixels classified.

Image analysts call the training data *ground truth* because field teams were often employed to identify pixels belonging to a particular class to use for training. Today, image analysts are more apt to use fine-spatial-resolution orthophotos to acquire the "ground" truth from image interpretation rather than field visits. However, some of these interpretations should be field checked to ensure accurate results.

The unsupervised classification approach

Analysts found that sometimes the groups of pixels that represented a particular class were not very similar spectrally. The "Urban" land use class is a good example. Pixels occurring in an urban area may be composed of concrete, asphalt, street trees, landscaping, and so on. All of these surfaces have different spectral signatures, and calculating an average spectral reflectance from the Urban class made little sense statistically because of the extremely high variance of such a class. If the spectral variance is high enough, almost any unknown pixel can be placed in the class with fairly high probability—a condition that causes excessive misclassification. The answer was to define numerous spectral classes through an *unsupervised* approach that would be later combined to make up the Urban land use class.

You can use mathematical clustering procedures to classify images in an unsupervised classification approach. These algorithms search for statistically separable classes in the input data, based on the spectral signatures of the pixels and criteria established by the analyst. When used alone to classify the image, clustering algorithms are called *unsupervised classification methods* because of the lack of analyst involvement once the algorithm is activated to process a dataset. The drawback to using clustering algorithms to classify an image is that none of the resulting classes have labels that describe the land cover information contained in the class. Also, and more important, although the classes fit the patterns of brightness of the multispectral data, they do not necessarily represent the information (thematic) classes desired for the mapping project.

As a result, clustering is more often used as a tool to create several spectral classes for heterogeneous land use categories, such as Urban, for example. The pixels identified as Urban are clustered into many spectral classes that have much lower variance than would be produced by forcing those pixels into one class. Once created and labeled, those spectral classes, developed by clustering, are included in the training dataset. Then, the maximum likelihood algorithm is used to classify the entire image. The training is both supervised and unsupervised in this approach, called *modified clustering* or *guided clustering* by Michael Fleming and Roger Hoffer at Purdue University in the 1970s. This approach has been used by many image analysts with slight modifications ever since (Lillesand, Kiefer, and Chipman 2008). GIS software generally does not provide the spectral class editing functionality necessary to combine supervised and unsupervised training in a guided clustering technique. Therefore, image classification is more effectively implemented with image processing software by image analysts.

Modern classification methods

At the turn of the millennium, the spatial resolution of satellite images improved by an order of magnitude with the introduction of two commercial Earth resources satellites: IKONOS and

QuickBird, reaching 30-centimeter resolution in 2014 with the WorldView satellite (chapter 4). This improved resolution means that per-pixel classifiers are less effective for producing land cover classification maps as they were when pixels were larger. The problem is that the pixels are so small that they do not represent a land cover class in general but rather spectral variations within a class, as explained previously in this chapter. Also, per-pixel classifiers can only assign unknown pixels to classes based on their spectral signatures alone without regard to the spectral signature of the neighboring pixels. Per-pixel classifiers therefore fail to consider the other elements of image interpretation (such as size, shape, and association of contiguous groups of pixels) that have long been recognized as helpful for interpreting and mapping land cover and land use manually from remote sensing imagery.

A new approach to image classification based on image objects that considers groups of pixels or regions of the image has been developed to address these issues. In this approach, groups of contiguous pixels are referred to as *segments*, and the algorithms that create them are called *segmentation algorithms*. The regions of pixels are also called *objects*, and the process that considers objects in images of the Earth is called *object-based image analysis* (OBIA; Blaschke, Lang, and Hay 2008).

Once the segments are identified, they must be classified. OBIA is approached in two steps. The image is first segmented into regions based on rules that can be modified by the analyst. The resulting regions are essentially machine-created polygons. They are delineated comprehensively throughout an area of interest. The process is repeated with different rule values until the size of the regions encompass the typical features evident in the image without including excessive variation within a region. The algorithms search for contiguous pixels having similar spectral brightness values. Normally, the process creates more regions (polygons) than an analyst would draw by hand, and those polygons have more-detailed edges and convoluted shapes than those drawn by hand (figure 8-3). Thus, OBIA software includes algorithms to smooth segment boundaries and eliminate very small segments.

Once created, the regions are classified by their spectral attributes, as well as their physical attributes of shape and size and their relationship to their neighbors. Some software packages, such as Textron System's Feature Analyst, combine the segmentation and classification of the image into one step. Other software packages, such Trimble's eCognition, separate the two.

Object-oriented approaches to automated image interpretation will likely replace per-pixel classifications when spatial resolution is fine (less than 5 meters). They have also been used on course-spatial-resolution images (1 kilometer) with success. Software engineers will likely continue the practice of creating add-on packages that connect to and work within GIS software to classify images. As a result, GIS analysts will be able to exploit these techniques without having to become familiar with a separate stand-alone software package. GIS analysts will need to understand the intricacies of image segmentation and classification, however, which is a significant undertaking. The accuracy of maps produced by object-based methods tends to be 5–10 percent better than those produced by pixel-based methods (approximately 90 percent correct rather than 80 or 85 percent correct), depending on the classes mapped.

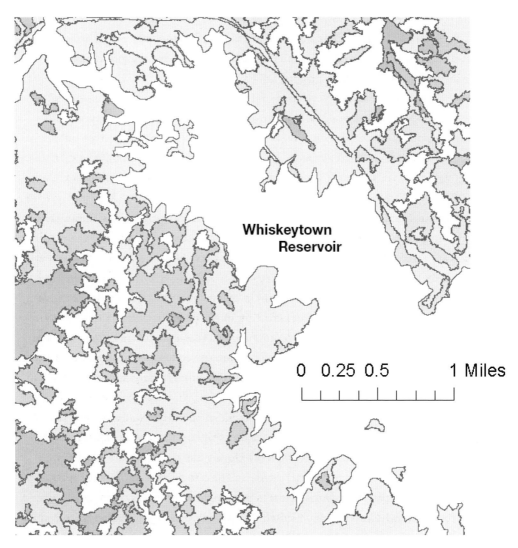

Figure 8-3. A portion of a vegetation classification map created with object-based image analysis (OBIA) methods, Whiskeytown National Recreation Area.

Map production by author based on data from McGovern et al. (2004).

Essential Earth Imaging for GIS

OBIA addresses a long-held criticism of image classification—that it produces a salt-and-pepper smattering of classified pixels rather than landscape units of similar attributes as traditionally defined by manual interpretation. OBIA creates polygons that can be identified individually rather than simply classified as pixels. The ability to identify landscape units has obvious advantages over pixel-based classifications. For example, a set of 15 lake polygons that can be attributed with names is a superior product to a set of 10,000 pixels classified as water. OBIA will likely dominate automated information extraction in the future because it better mimics human interpretation and produces more accurate classifications than pixel-based methods, especially with fine-spatial-resolution images.

One especially encouraging development is the increasing realization that the results of OBIA can be edited by hand more easily than pixel classifications. That is, the number of erroneously classified pixels always exceeds the number of erroneously classified polygons, because pixels are much smaller than polygons. Thus, the subset of erroneously classified polygons can be more easily reclassified or incorporated into adjacent polygons than the much larger subset of misclassified pixels. Rather than replace manual image interpretation with automated feature extraction, an optimum approach automates first and then edits manually (see the Equinox Analytics website, http://www.eqxanalytics.com). For example, most of the errors associated with automated delineation of buildings are associated with shadows from the buildings. These shadows occur at predictable places and can be more easily found and corrected with polygon-based OBIA classification software than with raster-based image classification. Also, the OBIA software more accurately extracts the buildings before manual editing.

Researchers looking to the future of extracting information from imagery are developing a more deterministic approach to object identification in which multiple reflectance values will be compared with library values to identify surface features automatically. And, in fact, this is occurring within the research community as advanced sensors measure an increasing number of wavelength bands (approximately 200 spectral bands with hyperspectral sensors), essentially providing spectroscopy from the air. Research is ongoing in tropical and arid environments combining hyperspectral imagery with lidar sensors (Carnegie Airborne Observatory 2014). The results of these advanced processing methods will continue to be delivered as data layers and three-dimensional point clouds that are compatible with GIS. Advanced image processing functionality will likely continue to require specialized image processing software, but some of that software will be integrated with GIS as an add-on product rather than a stand-alone software suite. This will allow GIS professionals to do more image analysis within a GIS software environment.

References

American Society of Photogrammetry. 1952. *Manual of Photogrammetry*. 2nd ed. Washington, DC: American Society of Photogrammetry.

Blaschke, T., S. Lang, and G. Hay, eds. 2008. *Object-Based Image Analysis: Spatial Concepts for Knowledge-Driven Remote Sensing Applications*. Berlin: Springer-Verlag.

California Department of Fish and Wildlife. 2014. "Biogeographic Data Branch Map Services." Accessed March 31, 2015. http://www.dfg.ca.gov/biogeodata/gis/map_services.asp.

Campbell, J. B., and R. H. Wynne. 2011. *Introduction to Remote Sensing*. 5th ed. New York: Guilford Press.

Carnegie Airborne Observatory. 2014. "Carnegie Airborne Observatory homepage." Accessed March 31, 2015. http://cao.stanford.edu/?page=home.

Colwell, R. N. 1954. "A Systematic Analysis of Some Factors Affecting Photographic Interpretation." *Photogrammetric Engineering and Remote Sensing* 20 (3): 433–54.

_____, ed. 1960. *Manual of Photographic Interpretation*. Washington, DC: American Society of Photogrammetry.

Esri. 2014a. "Esri Products." Accessed March 31, 2015. http://www.esri.com/products.

_____. 2014b. "ArcGIS Online Image Services." Accessed March 31, 2015. http://www.esri.com/software/arcgis/arcgisonline/maps/image-services.

Fox, L., III, and R. L. Garrett. 2002. "A Wildlife Habitat Map and Database for the ORCA (Oregon-California) Klamath Bioregion Derived from Landsat Imagery." In *Proceedings of the Fifth International Symposium of the Use of Remote Sensing Data in Mountain Cartography*, 27. Karlstad, Sweden: Karlstad University Studies.

GIM International. July 2011. "Operational SAR Tools for ArcGIS." *GIM International*. Nieuws feature article. http://www.gim-international.com/news/id5470-Operational_SAR_Tools_for_ArcGIS.html.

_____. 2012. Feature articles and editorial. *GIM International* 26 (12): 13–31.

GRASS GIS. 2014. "GRASS GIS homepage." Accessed March 31, 2015. http://grass.osgeo.org.

Iliffe, J., and R. Lott, eds. 2008. *Datums and Map Projections: For Remote Sensing, GIS, and Surveying*. 2nd ed. Boca Raton, FL: CRC Press.

Intergraph. 2014. "Intergraph Solutions." Accessed March 31, 2015. http://www.intergraph.com/sgi/default.aspx.

Jacobsen, K. 2009. "Geometry of Vertical and Oblique Image Combinations." In *Remote Sensing for a Changing Europe: Proceedings of the 28th Symposium of the European Association of Remote Sensing Laboratories, Istanbul, Turkey, 2–5 June 2008*. Amsterdam: IOS Press.

Liang, S. 2004. *Quantitative Remote Sensing of Land Surfaces*. New York: Wiley.

Lillesand, T. M., R. W. Kiefer, and J. W. Chipman. 2008. *Remote Sensing and Image Interpretation*. 6th ed. New York: Wiley.

Lu, D., and Q. Weng. 2007. "A Survey of Image Classification Methods and Techniques for Improving Classification Performance." *International Journal of Remote Sensing* 28 (5): 823–70.

Maalouli, G. A. 2007. "Aerial Perspective: Softcopy Photogrammetry." *Professional Surveyor Magazine* 27 (1). http://archives.profsurv.com/magazine/article.aspx?i=1770.

McGlone, J. C. 2013. *Manual of Photogrammetry.* 6th ed. Bethesda, MD: American Society of Photogrammetry and Remote Sensing.

McGovern, I. P., L. Fox III, S. J. Steinberg, and J. D. Stuart. 2004. "Utilizing High Resolution Imagery and Feature Extraction for Vegetation Mapping at the Whiskeytown National Recreation Area, Redding, California." *ASPRS Technical Papers*, CD-ROM: Year 2004 Annual Conference. Bethesda, MD: American Society for Photogrammetry and Remote Sensing.

NASA. 2013. "Landsat Ecosystem Disturbance Adaptive Processing System (LEDAPS)." Accessed March 31, 2015. http://ledaps.nascom.nasa.gov/index.html.

NASA/JPL. 2013. "Shuttle Radar Topography Mission." Accessed March 31, 2015. http://www2.jpl.nasa.gov/srtm/.

Olson, C. E., Jr. 1960. "Elements of Photographic Interpretation Common to Several Sensors." *Photogrammetric Engineering and Remote Sensing* 26 (4): 651–56.

Quattrochi, D. A., and J. C. Luvall, eds. 2004. *Thermal Remote Sensing in Land Surface Processes.* Boca Raton, FL: CRC Press.

Reeves, R. G., ed. 1975. *Manual of Remote Sens*ing. Bethesda, MD: American Society of Photogrammetry and Remote Sensing.

Renslow, M. S., ed. 2012. *Airborne Topographic Lidar Manual*. Bethesda, MD: American Society of Photogrammetry and Remote Sensing.

Richards, M. A., J. A. Scheer, and W. A. Holm. 2010. *Principles of Modern Radar: Basic Principles*. Raleigh, NC: SciTech Publishing.

Slama, C. C., C. Theurer, and S. W. Henriksen, eds. 1980. *Manual of Photogrammetry*. 4th ed. Falls Church, VA: American Society of Photogrammetry.

Stoney, W. E. 2008. *ASPRS Guide to Land Imaging Satellites*. Falls Church, VA: American Society for Photogrammetry and Remote Sensing. Accessed March 31, 2015. http://www.asprs.org/a/news/satellites/ASPRS_DATABASE_021208.pdf.

Trimble. 2015. "Trimble UX5." Accessed March 31, 2015. http://uas.trimble.com/ux5.

US Geological Survey. 2007. "USGS Digital Spectral Library." Accessed March 31, 2015. http://speclab.cr.usgs.gov/spectral-lib.html.

Vosselman, G., and H.-G. Maas. 2010. *Airborne and Terrestrial Laser Scanning*. Boca Raton, FL: CRC Press.

Waldman, G. 1983. *Introduction to Light: The Physics of Light, Vision, and Color*. New York: Prentice-Hall.

Wang, F. 2012. CS 201: *High Performance Medical Imaging GIS*. Los Angeles: UCLA Engineering Computer Science Seminar. Accessed March 31, 2015. http://www.cs.ucla.edu/events/events-archive/2012/cs-201-high-performance-medical-imaging-gis-fusheng-wang-emory-university.

Williams, D., and D. Brown. 2006. *Survey of US Federal Government Departments and Academic Institutions that Perform Archival Services or Collect Satellite Imagery Data Products*. Falls Church, VA: American Society for Photogrammetry and Remote Sensing. Accessed March 31, 2015. http://www.asprs.org/a/news/satellites/Satellite_Archives_v3.pdf.

Wolf, P. R. 1983. *Elements of Photogramme*try. 2nd ed. New York: McGraw-Hill.

Index